THE PRIVATE LIVES OF BIRDS

BRIDGET STUTCHBURY

THE PRIVATE LIVES OF BIRDS

A Scientist Reveals the Intricacies of Avian Social Life

Walker & Company
New York

Published by Walker Publishing Company, Inc., New York

All papers used by Walker & Company are natural, recyclable products made
from wood grown in well-managed forests. The manufacturing processes conform
to the environmental regulations of the country of origin.

LIBRARY OF CONGRESS CATALOGING-IN-PUBLICATION
DATA HAS BEEN APPLIED FOR.

ISBN 978-0-8027-1746-7 (hardcover)

First published in Canada under the title *The Bird Detective* by
HarperCollins Publishers Ltd in 2010
First published in the United States by Walker & Company in 2010

1 3 5 7 9 10 8 6 4 2

Printed in the United States of America by Worldcolor Fairfield

I thought of my friends who never take walks . . .
"for there was nothing to see."
I was amazed and grieved at their blindness.
I longed to open their eyes to the wonders around them;
to persuade people to love and cherish nature.

—Margaret Morse Nice (1937),
pioneering ornithologist and author of
The Watcher at the Nest

CONTENTS

LIST OF ILLUSTRATIONS
BY JULIE ZICKEFOOSE

THE PRIVATE LIVES OF BIRDS

Introduction

The first time I hiked through the forest that is now my summer home, I was blind and deaf to the secret lives of its inhabitants. It was June 1990, and I was in the final months of my PhD studies at Yale University and ready for a change. After spending the previous five years studying how swallows fight for nest boxes I had decided that my next research project would be on a small forest bird. No more birdhouses and lawns for me! I was in northwestern Pennsylvania, visiting the Hemlock Hill forest that would be my study site the next spring and, though I didn't know it at the time, for many years to come. On the long drive I had dutifully studied my field guide and listened

1

to birdsong tapes in my pickup truck. I was eager for a glimpse of my new study species, the hooded warbler, which I had never before seen or heard. This handsome little bird is unmistakable—males have a black crown and bib that contrasts sharply with bright yellow cheeks and a yellow belly.

I heard plenty of birds singing in the forest and could pick out some of the familiar characters: the rose-breasted grosbeak who sounded like a crisper version of a robin, the wood thrush with its beautifully complex and resonating voice, and the chestnut-sided warbler who cheerfully greeted me at the forest edge with a raspy *pleased, pleased, pleased to meet cha.* Even though it was a bright, sunny morning, the forest floor was dark with shadows from the closed canopy of leaves overhead. Birding by ear is relaxing, but actually spotting the songsters involves some frustration and disappointment. I heard several renditions of the high-pitched hooded warbler song I had memorized, *weeta weeta weet-e-o,* but I saw only one bird, a female who flashed her white outer tail feathers several times while barking a metallic *chip* call of alarm, and then promptly disappeared into a tangle of wild raspberry.

When I returned to Hemlock Hill the next spring, I realized the forest was full of hooded warblers, and from one spot I could hear half a dozen males singing. Of course the warblers had been there the year before, but my untrained ears had not been able to pick them out among the chorus of other songs.

I've now spent twenty years hunting in this forest with my binoculars and notebook as a biology professor, and know the terrain and birds like the back of my hand. The soft *chiff* from the hemlocks by the stream is a female Acadian flycatcher, the nasal *yank* above my head is a blue-headed vireo that is near its nest or mate, and the *chee chee chewup chewup* is one of the many songs of the hooded warbler. Three years ago, the broken-off hemlock near the thick grapevine tangle halfway up the stream held a hidden treasure—a hard-to-find scarlet tanager nest.

My favourite forest bird is the scarlet tanager; males are a stunning, brilliant red with contrasting jet black wings. Despite their flashy colour and bold song, males become very quiet after pairing and are hard to see. Male tanagers seem to disappear into the canopy and spend most of their time near their mates. During nest building, the female flies to and from the nest and does virtually all the construction while the male sits quietly a short distance away, keeping a close eye on her.

I remember the first time I heard a soft, rising *pew, pew, pew* sound overhead and guessed that its owner was a vireo, grosbeak, or flycatcher (wrong on all counts). This subtle call of the female scarlet tanager is a magnet for the male, who upon hearing it promptly flies over to perch near her. This is a sound I learned only after spending many hours watching scarlet tanagers and, although scarlet tanagers are popular birds, this particular sound is not to be found on bird identification CDs. A friend who specializes in recording and producing such audio field guides remarked that he had no idea that scarlet tanagers even made these sounds. Yet most female tanagers communicate with their mates in such a manner, and this seemingly unremarkable sound of the forest is rich with meaning and consequences for the intended audience.

Males are especially attentive to their mates during incubation because they deliver food to the female. Male tanagers are usually quiet when near the nest, but it is hard to be secretive when you are the colour of a stop sign. The male does not boldly announce his presence but instead perches about 20 metres from the nest with a nice caterpillar in his beak, and sings his song very, very quietly. I call this "whisper song"; it has a ventriloquial effect, and I know I am not imagining things because I can see his beak opening and closing slightly in rhythm to his throaty robin-like song. The female, who is on her nest, looks over at him and starts quivering her wings and making a loud, grating, begging call that sounds much like that of a large nestling. Watching the bright red male fly right to the nest, and listening to the female's loud calls, I can't help wondering how this nest could escape the attention of predators like blue jays and crows. Other female tanagers are more cautious, and fly off the nest to get food from the gaudy male.

To find out why male scarlet tanagers are so preoccupied with feeding their mates, I recently conducted a study with my colleague Paul Klatt, who is a biology professor at Ferris State University in Michigan. Paul spent three summers at the Hemlock Hill forest relaxing on a reclined camouflage-patterned lawn chair and peering at tanager nests high overhead in the forest canopy. He wrote all his notes in a small spiral-bound notebook: "7:23 a.m. Female off the nest, flew to the male, begging, male fed. Pair foraged together quietly, within 15 m of each; 7:29 a.m. Male fed female; 7:35 a.m. Female back on nest."

Females who were fed often by their mates spent more time on the nest incubating, which suggests that females gain nutritionally from the male's help. But we also found that the extra time on the nest did not translate into higher nest survival, so by the end of summer females that had received little aid had produced just as many broods as well-fed females. We were puzzled why males worked so hard to feed their mates if, in the end, this did not benefit the male by increasing his nesting success.

This domestic scene could be interpreted easily as simple parental co-operation; it's hard work raising a family and this job

4

is better accomplished with the efforts of both parents. Only the female incubates the eggs, and her breaks from incubation duties are short because the eggs have to stay warm; male food deliveries help to make up for the limited time she has for feeding herself. There is more to the story, however, because the social lives of birds are also driven by selfishness, conflict, and competition.

Paul and I wanted to know what the female would do if the male did not come with his little treats. After watching a pair for several hours under normal conditions, we caught the male and I took him to a covered birdcage far off the territory. Meanwhile, Paul watched the female intently to gauge her reaction. The female typically sat on her nest incubating the eggs, as expected, but remained there twice as long as normal, as though waiting for her mate to arrive with food. When the female finally did leave her nest, often after an hour of waiting, she began giving harsh alarm calls, *chik burr,* over and over again. Some females called loudly for over twenty minutes before returning grudgingly to the nest. Under normal conditions, when we weren't doing experiments, females rarely gave alarm calls, and when they did the male quickly arrived on the scene.

We found out the hard way why a male cannot ignore his mate's angry cries. During one experiment the female was especially upset and had been off her nest alarm-calling for almost an hour, so we decided to release the male. I was just walking over to get him from his cage when the female stopped calling, looked one way and then the other, and abruptly flew off her territory. I ran and got the male and let him go within a few minutes, and he obligingly began singing. Paul watched the nest for another hour, and the next day, but the female never came back. The male had lost his mate because he had been gone for an entire hour; to the female this was clearly unacceptable.

I was stunned that a female would abandon healthy eggs just because the male was a little late with the caterpillars. A female routinely catches plenty of food by herself, but the crunch time comes when the eggs hatch and the four chicks need to be fed many times an hour for three to four weeks before they can take care of themselves. A female depends on her mate's ability to find food; if a male fails the test during incubation it may be better for the female to cut her losses and start over again elsewhere with a new male. A female is not simply a passive recipient of a free handout, but coerces her mate into bringing a steady supply of food as a test of his parenting skills.

The job of the scientist is to identify interesting questions and find the answers by gathering convincing evidence, something that casual observation cannot usually accomplish. Hypothesis testing is the foundation of the scientific method, and the most difficult part of doing field studies on birds is collecting the right kind of information, in large enough quantities, to test one idea against another. How do you observe birds, find and monitor their nests, determine who mates with whom, and keep track of survival? It is not always easy to catch and band birds, or even to see what they are doing. Scarlet tanagers are not domesticated in the slightest, are difficult to lure into a net, and build high nests that are often well concealed by the forest canopy. It takes us four to five hours, on average, to find a single nest and even longer to catch the pair who lives there. This kind of painstaking work is needed to explain the most basic questions about bird behaviour and to reveal the incredibly complex social lives that most birds lead.

The obvious answer to why male birds are often flashy and ornamented is so they can impress females, but this barely scratches the surface. Females, because they make protein-rich and energetically expensive eggs, are naturally very picky in their choice of mates. Females use every trick in the book to judge a male's worth as genetic contributor to her offspring and resource provider for her family. Female purple martins in the United States prefer older males, female widowbirds in Africa prefer males with longer tails, and female blue tits in Europe prefer males with a varied song.

A good starting question, "Why do birds sing?" soon leads to wondering why some birds have a special tune they sing only in the darkness before dawn and why copying a rival's song is so effective for intimidating him. Birdsong is music to our ears, but to birds the love song is also a sophisticated weapon to keep competitors at bay. Territory ownership is advertised to neighbours and guarantees the private use of important resources like nest sites, mates, and a steady food supply. Though we may not think of birds as being particularly violent, intruders who do not heed the warning songs are attacked and may suffer serious injury in the brawl.

There are many fascinating stories hidden in the melodies of the robin, the flash of orange on the redstart, and the male tanager who feeds his incubating mate. Around the globe, competition for sex and resources is near universal. Male birds are under pressure to impress mates, females hold out for the highest-quality males, parents must share the burden of child care, and neighbours fight over space and food. The details of how these sexual and personal conflicts are resolved vary among species, and lead to surprisingly rich and complex private lives in birds.

~◡

A year after my first visit to the Hemlock Hill forest, I became a professor at York University in Toronto, Canada, which is only a four-hour drive away. My research on the secret lives of birds is a family affair, as my husband, Gene Morton, is also an ornithologist, and our children come with us on our expeditions. Gene worked for the Smithsonian Institution in Washington, D.C., for many years; his specialty is deciphering birdsong. My early detective work as a graduate student took me from the fields of Ontario to the deserts of Arizona to the tropical forests of Mexico and Brazil. When our children were babies and toddlers, Gene and I spent several months each winter living in Gamboa, Panama, where we studied tropical rainforest birds.

My children gleefully leave school in Toronto a month early so that we can move to our summer home in Pennsylvania in late May, when the nesting season is in full swing. Douglas's and Sarah's summer activities include soccer, bicycling, and swimming, but they also help their mom put up nets for catching birds, peek into nests to count eggs, and band baby birds before returning them to their cozy nests. The forest behind our old farmhouse is home to thousands of songbirds and makes a perfect "laboratory" for our research. Over the past two decades, I have conducted studies on many of our feathered neighbours, including the hooded warbler, scarlet tanager, wood thrush, and Acadian flycatcher.

In the fall and winter my family lives in the suburbs of Toronto, near York University. My morning routine is typical of that of many suburban mothers: have a cup of coffee, walk the dog, make my lunch, and bustle around frantically getting my kids ready for school while eyeing the clock to be sure they don't miss their bus.

One morning Sarah called from the basement, "Mom, can we give Gary a bath?" It was already 7:50 a.m. and she was still in pyjamas.

I replied, "Can't we do this *after* school?" Sarah is what one would politely call determined, a trait she got a double dose of from her parents.

"But, Mom, I'm afraid Gary will get sick if we don't give him a bath *now.*"

She had a point, but nevertheless I stalled.

"Have you finished your homework? Have you got your lunch ready? Have you brushed your teeth?" She answered "yes" to all of my questions, so I went downstairs with a bucket.

Gary is Sarah's pet garter snake. As I stood there watching Gary swim around in the shallow water, with my nine-year-old daughter beaming like a proud parent, it occurred to me that I must be the only mother on my street, probably in the world, at that moment, giving a snake a bath. Yet this morning detour did not strike me as particularly odd, given that there are many everyday events in my life that, I have to confess, are a bit unusual. My field studies on birds have found me quietly stalking a subject through the forest for hours at a time, delicately taking blood samples from an outstretched wing, and hacking my way through a dense tropical forest with a machete to make room for my long bird nets.

I've spent the better part of my career as a biology professor mounting miniature radio-tracking devices on songbirds so I can study how and why they cheat on their mates. I follow the philanderers through the forest as they sneak off to have a one-minute stand with the next-door neighbour. I started my scientific career as a behavioural ecologist, studying what I think of as the fun questions about the evolution of bird behaviour. One of my first scientific papers, published in 1987, was titled "Spitefulness, Altruism, and the Cost of Aggression: Evidence Against Super-territoriality in Tree Swallows." This was followed the next year by the more ominous "Experimental Evidence for Sexually Selected Infanticide in Tree Swallows." At that time, the buzzwords of *conservation biology, biodiversity,* and *climate change* were in their infancy.

Today, only twenty years later, many ecologists have turned their attention to triage and finding cures for imminent extinction and the enormous biodiversity loss that is underway. Birds are highly sensitive to pollution, habitat loss, and climate change. Around the globe, bird populations are in sharp decline and experts forecast a shocking but realistic 15-percent extinction rate within the next century. This amounts to the loss of over one thousand species of birds in my children's lifetime.

Many of my most recent studies are very practical, for instance, measuring the survival of juvenile songbirds in forest fragments and finding out how tropical deforestation affects the health of wintering migrants. But I am still up to my old tricks and recently published the study on mate feeding by male scarlet tanagers and another one on cuckoldry in the wood thrush. What is the connection? Behaviour, and the ability to adapt to new circumstances, is what could make or break a species' future survival. As spring arrives earlier each year, can birds simply learn to begin breeding earlier? When the last forests on a tropical island are cleared, can birds adopt a new lifestyle among the scrubby fields and pastures? Understanding the past evolution of bird behaviour is critical for predicting how the future will unfold.

How does a species adapt its behaviour to a new environment? There are two fundamentally different ways: learning by individuals, and the spreading of beneficial genes in a population. The pace at which these two adaptive mechanisms occur is dramatically different. New behaviours that arise in a population via learning can appear suddenly and spread through the population within a few generations. In contrast, individuals with a new, beneficial gene will enjoy a subtle but innovative advantage over others who do not carry the mutation, and will pass on the new trait to their children, and so on. Behaviours that require changes at the genetic level often take hundreds, if not thousands of years to become common in the population.

A case in point is the classic example of blue tits in Britain in the 1950s. They learned how to poke holes in the foil caps of the milk bottles left on the doorstep, to drink the cream at the top. This innovative behaviour, which led to a big food reward, was quickly copied by other blue tits. There was no gene coding for "milk-bottle opening" but rather an adventurous individual discovered the joys of cream through its curious and bold behaviour and its built-in tendency to explore unusual environments and foods. This is not to say that milk-bottle opening is likely to arise in any species of bird or that it has no genetic basis whatsoever. The ability to learn, the tendency to explore new habitats and foods, and the tendency to copy are all behaviours that *do* have some genetic basis and differ greatly between species and even individuals.

Imagine if we had to wait for a mutation for milk-bottle opening behaviour to arise completely by chance as a result of mistakes in DNA copying inside the cell. Mutations happen often inside our germ cells, but most have no effect on our offspring because the mutation occurs in a part of our DNA that actually has no job, the so-called non-coding regions. When a mutation happens in the sections of DNA that make protcins, however, this usually interferes with the complex process of reading DNA sequences to make the right amounts of the right proteins, at the right times. A popular analogy is randomly removing a component of a computer and expecting it to work better as a result. Only rarely is a mutation actually beneficial to the offspring who inherit it.

Even if a beneficial mutation arises by chance, it may not spread rapidly through the population. First, the young birds carrying the new mutation may die for reasons completely unconnected to their genetic makeup, for instance, if a hungry crow eats them before they leave the nest. The long wait for the same accidental mutation would have to start over. Second, most birds have a distinct breeding season and a parent produces only one, or less commonly two, small batches of young per year. Scarlet tanagers, for instance, breed only from May to July and each pair produces at most three to four nestlings during the short breeding season. These offspring have to survive their long migration journey to the tropical rainforests of South America, where they spend the winter before returning to breed (and pass on their genes) the next summer. Birds like gulls, shorebirds, hawks, and albatrosses take several years to become sexually mature, so it takes even longer for an advantageous mutation to spread from one generation to the next.

I wonder how, and if, birds can adjust to the rapidly changing world they live in today. In just a few centuries, the native grasslands and towering hemlock and oak forests that greeted the pioneers to North America have been replaced by hayfields and rows of corn. Millions of hectares of tropical forests have been turned into pastures, banana plantations, and soybean fields. The natural forces that have shaped bird behaviour for thousands of years have shifted as seasons change, food supplies crash, and habitats disappear. Behaviours that once paid high dividends may today doom an individual, and a species, to failure. Birds are linked together, and to us, by the challenges of living in the drastically remodelled environment that has come with explosive human population growth. Birds connect people with nature because they are beautiful, have fascinating lives, and in many ways remind us of ourselves.

I am a bird detective, revealing the behind-the-scenes details of the social lives of birds to understand why females cheat on their mates, what makes a male attractive, why some pairs divorce, how birds claim a territory—and what all this means not only for our avian friends, but for us as well.

1 PHILANDERING FLYCATCHERS
Why Females Cheat on Their Mates

There was a fresh coating of icy frost on the grass, glistening in the porch light as I quietly snuck out of the house before dawn. I left a trail of ghostly footprints behind me, marking where my warm boots melted the frost, but even these soon disappeared. In daylight, it was an easy ten-minute walk through the forest to the second stream, but in the early morning darkness, even with a flashlight, I had to pick my way slowly along the path. The day before, I had marked trees along my route with reflective tape, and now I paused often to scan ahead as I stumbled from one gleaming signpost to the next.

As I crossed the farm stream, the one that feeds our pond, I heard a wood peewee let loose a loud, mournful *peeee-eee weee;* this flycatcher is one of the first birds to sing in the pre-dawn hours and I knew I had better hurry because the forest was awakening. I arrived out of breath, with puffs of steam swirling out of my mouth in the beam of the flashlight. Then I abruptly cast myself into darkness again. Though I could see the first light of dawn through peepholes in the canopy, it was still dark on the forest floor.

The forest began bursting into life. Nearby, a scarlet tanager sang his raspy version of a robin song punctuated with a *chick-burr* call note, a second peewee sang overhead, and that was followed by the hauntingly beautiful song of the wood thrush, *ee-o-lay-ay-ay.* The male Acadian flycatcher that lived by the second stream was invisible but not far away, singing a complex, sputtering song heard only at dawn. Acadian flycatchers prefer dark stream valleys within the forest and are well camouflaged among the hemlocks on account of their bland green-olive plumage that completely lacks ornamentation.

I took off my gloves and fiddled with a small radio attached to my belt until I heard rhythmic *beeps.* About 40 metres away, a female Acadian flycatcher was carrying a tiny transmitter on a backpack that sent out a radio signal, giving away her location to anyone listening to that particular frequency. I held what looked something like an old television antenna, small enough to grip in one hand. I aimed the antenna into the darkness and slowly spun it in a full 360-degree circle. The beeps were loudest over near the edge of the stream, confirming that the female was on her territory and just south of the nest I had seen her building a few days ago.

I followed the female from a polite distance as the trees slowly took on their shapes, and I paused to write times and locations in my notebook. She stayed high in the hemlock trees, and without the radio-transmitter on her back it would have been impossible for me to know her whereabouts.

The reason I was here was to find out if female Acadian flycatchers roamed off-territory to visit neighbouring males at dawn, like some other birds are known to do. I had good reason to believe female Acadians would cheat on their mates. Our DNA testing over the past few years had shown that 41 percent of the Acadian flycatcher nestlings in this forest were the result of a female's eggs being fertilized with sperm from a male other than her mate. But who was responsible for arranging the rendezvous, the male or the female?

Over the course of the next two hours my quarry did not stray far from her nest as she hunted for her breakfast; she needed extra food to make the eggs that were still hidden inside her body. By 7 a.m., I was chilled to the bone and a touch disappointed at the female's lack of philandering. I was beginning to realize that female Acadian flycatchers do not go sneaking off into the darkness through the forest—though certain female biologists do.

No one was awake in the farmhouse when I got home. Douglas and Sarah were in grades two and three at the time and were well accustomed to their mom's unusual job. On this morning in late May, I had rushed out before dawn to track the female flycatcher because she was at her peak fertility and would lay eggs any day. I couldn't afford to miss a day and had had just enough time to fit in a two-hour radio-tracking session before my family left to visit grandma in Cleveland. I took off my wet boots and put away my tracking equipment, then switched gears and started packing the sandwiches, drinks, and colouring books.

In all, my students and I radio-tracked a dozen different female flycatchers adding up to over one hundred hours of detective work, but not a single female ever left her territory. The apparent monogamy, a female living with one male and the two raising a family together, would have been convincing had it not been for our genetic evidence of frequent infidelity.

David Lack, a famous ornithologist who developed many key ideas in the 1950s, proclaimed that 90 percent of birds worldwide are monogamous; he thought this because they form monogamous pair bonds. One of the avian world's best-kept secrets was the stunningly high level of infidelity in many types of birds, revealed only when DNA testing became possible. We have learned to distinguish carefully between social monogamy—the obvious pair bonds—and genetic monogamy, which reflects the well-hidden mating decisions of the female.

Ornithologists use the term *social* male to refer to the male with whom a female is paired; this male typically defends his shared territory and nest and cares for the young. The *extra-pair* male is a male who lives elsewhere, provides no care for the young, but who is actually the genetic father of at least one nestling in the female's nest. The most intense competition among males often happens after pairs are formed, when females choose who will father their eggs.

How do eggs actually get fertilized and why does this make a male's job so tricky? The timing of fertility, and sexual receptivity, is very different in female birds than in women. During the egg-laying period a female bird ovulates once a day, or every other day in some species, and releases an ovum that soon enters the egg-making tract, or oviduct. Fertilization of the egg can occur only during a short window of time, often thirty minutes, when the ovum is floating around deep inside the female's body.

Whether or not the ovum is fertilized by sperm, it travels down the oviduct and over the next twenty-four hours receives its various layers of egg white and eggshell, and pops out the other end. Our omelettes and devilled eggs are usually made with unfertilized, rooster-free, chicken eggs.

How, then, is a male to get his sperm to the top of the oviduct where it can lie in wait for a passing ovum? Male birds can transfer millions of sperm in a single copulation, but this alone does not guarantee success. Once a female begins egg-laying, most sperm swimming up the oviduct will get pushed back out by the large egg coming down the tube. In many birds, females have tiny side tunnels at the base of the oviduct where sperm can hide, and remain viable, for many days. The trick for the male, then, is to copulate with the female during the days leading up to egg-laying and hope to store sperm inside her, or to copulate in the minutes or hour prior to her ovulation.

If the female mates with multiple males, then the eggs within her nest may have different fathers. More sperm in the oviduct, particularly right before ovulation, means that a particular male has a numerical advantage over other males in the odds of fertilizing a given egg—like winning the lottery.

For practical reasons, a female bird is usually sexually receptive to the advances of males only when she actually needs sperm to fertilize eggs. This occurs during the days she is making and laying eggs plus about a week beforehand when sperm can be stored. Females who are incubating eggs or feeding nestlings usually ignore the advances of males. Males, on the other hand, produce sperm all season and are ready to mate at a moment's notice, but the hard part is finding an egg-laying female and then mating often enough, or at the right time, to fertilize an egg.

Finding out who was the genetic winner of the sperm competition in the Acadian flycatchers meant taking tiny blood samples from the nestlings and parents for paternity testing in my laboratory at the university. On a hot June afternoon I found myself inching my way up a ladder with a paper lunch bag firmly clenched in my teeth. I glanced up and saw there were six more rungs to go until I would be close enough to the nest to reach up and gently remove the small squirming nestlings. Acadian flycatchers build their nests in a fork near the end of a long branch, often with a long strip of grapevine bark dangling from the underside of the nest. From a distance, the nest looks like debris that caught on the branch as it fell from the canopy.

I could hear the female flycatcher angrily *chiff* at me and she occasionally hovered near her nest in alarm. When I reached the top of the ladder I slowly stretched out one hand and groped at the nest. I could feel a warm, soft mass, and gently pulled out the two nestlings. The female flycatcher dive-bombed my hand in a futile attempt to protect her babies from this large predator. It was impressive that a tiny 20-gram bird was willing to take on a 50-kilogram person. Of course, there was no way she could know that I was going to climb the ladder again in a few minutes to put her nestlings back unharmed. They were about ten days old, with half-grown feathers and alert eyes, and were the perfect age for banding and taking a blood sample. I popped them in the paper bag and started my descent.

To find a genetic match to the nestlings we also had to sample the DNA from the parents and from the other males in the forest who *could* be the father. Male Acadian flycatchers are reclusive forest birds, drab in colour, and are easily overlooked. They are, however, fiercely aggressive and chase away other males who dare to sing within the boundaries of their territory. We catch males by placing a small speaker underneath a special lightweight mist net and play an Acadian flycatcher song to trigger a testosterone-driven attack on the bold intruder. The net is 6 metres long and 2 metres high, and held up with long poles at either end that are pushed into the ground. A mist net, named for its wall of fine nylon threads, is rather like a large, rectangular, non-sticky spider web. To lure the male down to ground level and give him a bull's-eye, I perch a stuffed Acadian flycatcher decoy beside the net.

Back in May, I had caught the male who was caring for these two nestlings. Males were setting up territories then and could easily be fooled by my fake intruder. When he first heard the play-back tape, he immediately stopped singing and stealthily moved toward the source of the sound. I had spotted him in the dark branches of the hemlocks, flicking his wings in agitation and flying back and forth in the canopy. Every twenty to thirty seconds the male changed perches, gradually getting lower and lower as he silently zigzagged his way toward his enemy. I watched, holding my breath, as he perched low in a tree on the opposite side of the long net. He looked at the decoy, still flicking his wings, and suddenly attacked. He hit the decoy on the head with his beak, fluttered for a moment, and then flew into the net. He was harmlessly enveloped in the fine netting and was trapped until I ran up and gently untangled him. A few minutes later he was banded and back in the hemlocks, and I had my precious blood sample.

By the end of the breeding season, the hours of nest searching, ladder climbing, and netting produced a plastic box filled with vials of blood from the two dozen families of Acadian flycatchers from our forest. In my laboratory at York University, we extracted the DNA from the blood and used a DNA sequencing machine to determine each individual's unique genetic "fingerprint." The two nestlings sampled that June afternoon revealed that the female I had radio-tracked had not been faithful to her mate. The nestlings did not match the DNA of her social mate, the devoted parent who had fed the nestlings dozens of times a day for three to four weeks. The only male who was a perfect genetic match lived at the other end of the forest. This distant male had also sired all the eggs that his own female laid and so had produced a total of four offspring that summer compared to none at all for the cuckolded male.

Just over half the female Acadian flycatchers fertilized their eggs with sperm from their secret partners. Some males doubled their reproductive success as a result of mating with promiscuous females, while other males worked hard defending a territory and raising a family, but genetically speaking were a failure. Only one quarter of all illegitimate young were sired by a next-door neighbour whereas the other sires were from four to five territories away, about 500 metres away on average. This means that an Acadian flycatcher female, in theory, can choose among over twenty males as potential mating partners.

Our radio-tracking showed how these encounters happen; it is males who visit females, and not vice versa. Acadian flycatcher males are unusually energetic in their search for extra mates. My postdoctoral student Bonnie Woolfenden had put radio-trans-mitters on males so we could find out how often males went in search of willing females and who they visited. Mated males left their territory about once every two hours, were gone an average of ten minutes at a time, and wandered up to 1,500 metres from their home base (Figure 1.1). This finding was surprising. Males of other territorial songbirds usually limit their search to adjacent territories and focus their efforts on neighbours because females are more likely to accept copulations from familiar males

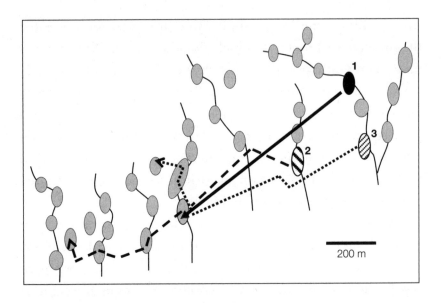

Figure 1.1. *This figure shows movements made by three radio-tracked male Acadian flycatchers in a Pennsylvania forest. Shaded ovals show the locations of nesting territories along streams (thin lines), and the movements of each male are represented by a different type of line (solid, dotted, dashed). All three males left their territories to visit distant females; one female was visited by all three. (After Woolfenden et al., 2005.)*

and because visiting close territories is less costly in time and energy to the male. When territories are packed close together, somewhat like a honeycomb, males have half a dozen females within easy reach.

Acadian flycatchers in our forest arrange their territories along streams, like beads on a string, so most males have only one or two females living beside them. A male who restricts his off-territory trips only to adjacent territories would have few opportunities to boost his reproductive success.

Male superb fairy-wrens in Australia might be thought of as particularly romantic because they impress females with elegant clothes and a gift of flowers. The trouble is they direct these displays to the girls next door rather than their own mates. Male breeding plumage is reminiscent of a flashy tuxedo, with the black breast and collar contrasting with a glossy pale blue crown, cheek, and nape. As soon as males have finished moulting from their dull brown non-breeding colours into their mating colours, they turn on the charm and visit females on neighbouring territories. The male perches in front of the female and shows off by spreading the feathers of his blue crown, cheek, and back in her direction. During many of these displays the male holds a bright yellow flower or petal in his beak.

It should come as no surprise to us that males sneak around; it is female behaviour that is so revealing. A visiting male must be patient because no copulation results immediately from his efforts. It will be months before the female is fertile and comes looking for him.

Andrew Cockburn, a professor at the Australian National University in Canberra, has studied superb fairy-wrens in the National Botanic Gardens since 1988. He found that over 60 percent of female fairy-wrens produce at least one offspring sired by a neighbouring male. Females have a good memory and keep track of when different males come calling. A male who moults into his breeding plumage early, and therefore begins his courtship displays earlier than other males, will be rewarded by having the female come under the cover of darkness, many months later, to visit him for a copulation. Females make their covert visits only during the pre-dawn hours and only during the few days of their fertile period when the copulation can fertilize one of her eggs.

Females care about the timing of male courtship visits because only males in good condition can afford the energetic expense of growing a new set of feathers when food is scarce in the austral winter months. This elaborate mating game allows a female to choose the best genetic father for her offspring regardless of whom she is paired with socially.

In European robins, intruding males offer females food not flowers. During the courtship period females beg loudly for food from their own mates using a far-carrying and high-pitched *seep* call that is quite similar to the begging calls of nestlings. Females need extra food for practical reasons and make more eggs if their mates feed them often. The begging call of a female is intended not just for her mate's ears, however, but for any male within earshot. The calls can easily be heard on neighbouring territories and a fertile female calls persistently, ten times per minute, even though her harried mate cannot possibly keep up with the demand.

Joe Tobias, at the time a PhD student at the University of Cambridge, temporarily removed males from their territory for an hour to find out what would happen if the male did not oblige the female. Females who were unfed by their mates almost doubled their rate of begging. Nearby males who were eavesdropping and heard the frantic begging calls of females were all too happy to pay a visit and trade food for sex. In essence, a female robin blackmails her mate into bringing a regular supply of food, with the very real threat of cuckoldry if he does not deliver.

Razorbills, chunky seabirds with short legs who look somewhat like penguins, cannot copulate standing up, unlike most birds. A female razorbill must co-operate by assuming a prone position, lifting her stubby tail and resting her chest on the ground. The male balances on her back and wraps his tail around and under hers so their respective genitalia come into contact. Male birds generally have the same minimal equipment as females, so he is at her mercy. The female can suddenly bring the encounter to an end by standing up, giving her near complete control over which male mates with her. Even a successful copulation lasts only sixty seconds. Richard Wagner, now a research professor at the Konrad Lorenz Institute in Austria, spent hundreds of hours during his PhD studies at Oxford sitting in a tiny blind on the rocks of Skomer Island, Wales, buffeted by the strong winds, hands numb from the cold, watching the sex lives of razorbills.

Razorbills build their nests under boulders to protect their eggs and young from predators. Each pair defends a tiny space around the nest site and when not standing on guard they are usually at sea diving for fish. Wagner noticed that groups of males lounged on a particular ledge and that fights broke out regularly, with males trying to push one another off. All the pushing and shoving was about females, of course. This ledge served as the equivalent of a singles bar, except that the males in attendance were already paired with a female and had a nest site somewhere under the boulders. Females that were already paired also visited the ledge and allowed attending males to mate with them. The mating arena was not, however, a free-for-all; only two males accounted for three-quarters of the copulations that Richard observed. There was nothing secret about these extramarital affairs; the acts took place in plain sight of all present, sometimes including a female's own mate, who stood watching helplessly.

~⌒

I recently gave a public lecture in Hanover, New Hampshire, near Dartmouth College, and my host cheerfully introduced me by telling the audience that "Bridget probably knows more about sex than anyone in this room." When lecturing in first-year biology classes, and even to upper-level students, my tales of infidelity in birds usually elicit giggles, elbowing, and knowing looks. One cannot easily discuss or explain animal behaviour without using terms that humans can relate to, and be amused by, because of the natural tendency to draw comparisons with our own social behaviour. There are many similarities in how competition and conflict have shaped the evolution of behaviour among animals, but we must not forget that birds do not have the same feelings, thoughts, or decision-making processes as humans. Here, I have tried to strike a balance between using detached scientific descriptions and terms that will strike a familiar chord.

To understand why females cheat on their mates we must understand the fundamental process of natural selection that drives the evolution of social behaviour in animals. The word *selection* refers to the winners and losers in life. In any population, some individuals will survive and reproduce better than others, and thus produce more offspring in a lifetime. Natural selection causes evolution only if the offspring inherit those traits that made their parents successful, in which case those genes become numerically more common as time goes on.

A female must therefore benefit from her infidelity via the DNA that her offspring inherit from the male who successfully copulates with her. A female's extra-pair partner usually contributes only sperm, because he does not come back to her territory to help her defend the nest or feed the young. A female will lay the same number of eggs no matter how many different males provided her with sperm, so having more partners does not result in an obvious increase in reproductive success.

Many aspects of a male's displays and physical prowess are under genetic control and some individuals carry beneficial mutations that give them an edge when competing for mates and resources, or staying alive. A male with a slightly larger body size may be better able to defend a territory, and his sons will enjoy the same advantage. A male may carry a mutation that helps him fight off a common disease, while others who lack the gene will fall victim.

Charles Darwin's most powerful evidence for natural selection came from the deliberate breeding of animals and plants by humans. Artificial selection occurs when humans arbitrarily choose the winners and losers in a population. We have created dachshunds and poodles from one ancestral species, and Brussels sprouts and cauliflower from another. Artificial selection experiments have shown both the genetic basis of bird behaviour and how quickly a population responds to strong selection pressure.

Great tits, common songbirds in Europe, are the subject of dozens of behavioural ecology studies. One team of researchers in the Netherlands found that some individuals tend to be risk-takers, while other individuals are more cautious. Starting with nestlings from a wild population, researchers tested the birds forty days after hatching using standardized lab tests. Birds were put into a novel environment, and the speed at which they explored their new habitat was measured. Next, they were put into a small cage and shown a novel object; the speed at which they approached the object, and how close they got, was measured. Finally, back in the aviary, the birds were deliberately startled when they came to a bird feeder, and researchers measured how quickly the birds resumed feeding.

Two breeding lines were established, one that bred "fast" birds together and the other that bred "slow" birds together. To avoid

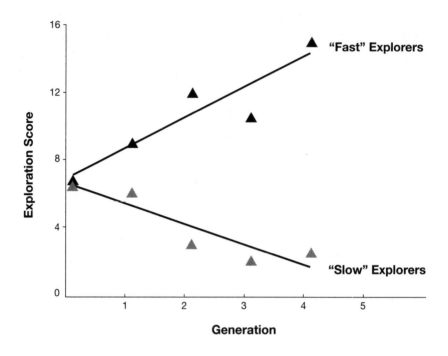

Figure 1.2. *This figure shows the effect of artificial selection on great tits' exploration scores. In each generation, individuals with similar exploration scores were paired in an aviary and allowed to produce young; only four generations later, the two populations were distinct. (After Drent et al., 2003.)*

the possibility that the behaviour was influenced by parental care, a pair's eggs were incubated by a randomly chosen foster mother. After hatching, the nestlings were mixed between fast and slow parents so there was no bias by learning behaviours from parents.

After only four generations of strong and consistent artificial selection on risk-taking, the two lines showed distinctly different behaviour (Figure 1.2). The positively selected line was 130 percent faster than the original wild population and the negatively selected line was 60 percent slower. The rapid evolution in this captive population occurred because of individual differences in risk-taking, a genetic basis for the behaviour, and strong selection pressure on that behaviour. Individual differences in the behaviour of wild birds has a genetic basis for such varied traits as the migration behaviour in European blackcaps, song learning in zebra finches, and even fear of humans in chickens.

Behavioural ecologists can rarely measure evolution directly in wild populations of birds, as it would take decades to show how the occurrence of a given trait changes over many generations. In many cases, we have not yet identified the gene(s) responsible for the behaviour. Instead, we measure natural selection in a single generation by studying if and how a behaviour is adaptive. In other words, we ask which traits (for example, high song rate, bright colours, large beak) result in the highest mating success, survival, and/or reproductive success for an individual.

Female choice can be a potent force in evolution if a handful of top-quality males in the population gain the lion's share of copulations and as a result their genetic traits become disproportionately common in the next generation. Female choosiness drives an arms race among males, and their ornaments and dance routines become ever more elaborate from one generation to the next as males compete for elusive copulations. As we have seen in fairy-wrens, razorbills, and Acadian flycatchers, even birds that are monogamously paired are not excused from the intense sexual competition.

⌁

There are two main theories as to why females not only accept copulations from neighbouring males, but in many species go out of their way to encourage them. The first is that females benefit from mating with males who possess "good genes" that will be inherited by their offspring. This appears to be the case in fairy-wrens and razorbills, where females judge a male's intrinsic quality by observing displays and fights. Evidence from other species shows that extra-pair young are healthier than their step-siblings. Blue tit nestlings sired by a neighbouring male, rather than the social father, were more likely to survive as nestlings. Similarly, collared flycatcher nestlings from the same nest fledged in better condition if they were sired by a neighbour rather than the male who was caring for them. Female mate choice is subconscious, driven by the successes and failures of many generations before them, and choices are made instinctively based on the traits most likely to indicate male quality.

The second theory is that females are attracted to mates whose genes are a good match to her own genetic makeup, so a male who is highly sought-after by one female may be ignored by another. In this case, there is no single "good gene" that females are seeking for their offspring, but rather they are looking for a male who is the most dissimilar genetically. Dissimilarity in the genes of parents can improve an offspring's ability to fight disease because of how the immune system works.

One of the most important families of immune system genes goes by the long name of *major histocompatibility complex,* or MHC. The MHC family of genes is found in all vertebrates; each gene is super-variable and made up of hundreds of different varieties. Each gene variant codes for a unique protein whose job it is to recognize and trap specific invading viruses and bacteria that are later killed by white blood cells. More types of MHC proteins manufactured by one's DNA means that more types of invaders can be intercepted and stopped in their tracks before the infection gets serious. A nestling with genetically different parents will inherit more varieties of the MHC and so will have a stronger immune system.

This genetic compatibility hypothesis has been tested in the Savannah sparrow, a grassland bird. Savannah sparrows are unremarkable birds, at least on the surface. They are in the category of "little brown jobs," small, dull-coloured sparrows that the average person would not look at twice. They do have an attractive yellow stripe over the eye, but even this is a relatively subtle decoration. Their scientific name, *Passerculus sandwichensis,* does not refer to some unusual local recipe from days gone by but rather to the Fourth Earl of Sandwich, who is renowned for his culinary invention using bread and who also underwrote many explorations of the New World. The voice of the Savannah sparrow is not very ear-catching, at least compared with a robin's, and consists of a long, high-pitched, buzzy song. The sight of males teetering on long grass stems as they sing is a common sight throughout the fields of northern North America.

One would not guess by their ordinary appearance that these sparrows have a complicated sex life. About half the males are bigamists, and have two females that nest on their territory. Why would females settle for sharing a male and his territory? These males control high-quality habitat where there is an unusually high abundance of insects, so both mates have more to eat and to feed their young than if they were the sole partner of a male who owned a lesser territory. Females arrive in spring on a staggered schedule, so female number one may be too busy nesting to prevent female number two from moving in.

Superimposed on these living arrangements is the now-familiar cuckoldry that occurs behind the scenes. Though about two-thirds of female Savannah sparrows produce illegitimate young, their choice of genetic mate is not based on his intrinsic quality (for example, good genes) but rather how genetically different the male is to the female. The young produced from genetically mixed marriages grow up in better condition than young produced from pairings where mom and dad were genetically similar.

But how do females judge a male's genetic compatibility? In mice, and humans, this is accomplished via the male's odour. Most birds have a relatively poor sense of smell, but olfactory cues have not yet been ruled out. Another possibility is so-called cryptic mate choice, an obscure way of saying that the male's success is determined unconsciously inside the female's body. Sperm from different males compete to fertilize the female's egg, and there may be some physiological mechanism that gives genetically dissimilar sperm an advantage in surviving inside the oviduct or beating the other sperm to the egg.

The complex mating system of the Savannah sparrow has arisen gradually over thousands of years, as a result of the uneven food resources, the staggered timing of female nesting, and the advantages of trading up the genetic variability of the young via extra-pair mating. Noah Perlut, as a PhD student at the University of Vermont, used mowed fields as an experiment to test how physical resources and timing affects the mating behaviour of Savannah sparrows. The new and increasingly common practice of mowing hayfields in early June, to increase the number of harvests per season, simultaneously kills almost all the nests in the field and causes females to re-nest en masse. Mowing also creates a uniform habitat, so males suddenly find themselves on an even playing field and females can no longer group up on luxurious territories.

The winners of sexual competition were fundamentally different in mowed fields, and whatever gains had been made before the mowing were violently erased. Perlut found that after mowing, larger males no longer held better territories, males had only one social mate, and cuckoldry became rare. By season's end, males with a large body size did not sire more offspring than ordinary males. The low rate of cuckoldry meant that many females could no longer obtain genetically compatible mates, which presumably compromised the future survival prospects for their young. About 40 percent of the hayfields in the Champlain Valley of Vermont are mowed early each summer, artificially changing the delicate interplay of male competition for mates and, by lowering nesting success, also driving population declines of the species.

Five hundred and eighty-three. This is how many times the female Acadian flycatcher had given her *chiff* call during the two hours since I began following her. The hatch marks on the pages of my field book were the proof—I had counted every single one of her repetitive calls, wishing that she would stop or at least take more than a two-minute break. In mid-morning the forest floor was bathed in sunlight; it was early spring and the leaves on the oaks, beeches, and maples had not yet closed a curtain over the forest. I certainly had heard the female flycatcher far more than I had seen her. Standing among the evergreen hemlock trees in a small stream valley, I could hear her calling overhead but only occasionally did I catch a glimpse of her as she darted out from a perch to catch small, flying insects.

This particularly noisy female was the fourth that I had fitted with a radio-transmitter and it was becoming obvious that I was not going to get much exercise that summer. My twenty hours of radio-tracking thus far had been spent standing more or less in one spot, eagerly pointing my antenna toward the hidden female, waiting for her to make her sudden move off-territory. I love being in the woods for hours at a time, drinking in the nature that surrounds me, but even I had to confess to having become a little bored with "following" these stationary females.

Brief moments of excitement sometimes occurred during *chiff* bouts when I saw another flycatcher dart toward the female and chase her aggressively for ten to twenty seconds. It was impossible to keep track of who was who amid the blur of feathers dodging between the branches. At least three times the aggressor was the female's own partner, who clamped onto her back, drove her to the ground, and attempted to mate with her. More often, there were three birds involved and after the melee one of them would sit quietly on a perch and sing *peet-za!* This was the owner of the territory, the female's mate, who apparently had just interrupted a sneaky copulation attempt.

Acadian flycatchers are one of the first birds in the forest to sing during the dawn chorus, when it is still dark. Males perform a distinctive dawn song display that can be heard several territories away. They sing a complex series of notes (*seet, tee-chup, speet*) very rapidly, giving about one call per second, and often keep this up continuously for over thirty minutes. We do not yet know if females listen to this vocal marathon and later are more likely to accept copulations from males with the best performance. Another possibility is that females are not looking simply for males with good genes, but rather select males who are genetically compatible, as is the case for Savannah sparrows. The close physical encounters that occur frequently on her own territory may give her all the information she needs.

Females who are fertile give *chiff* calls four times more often than females who have already laid eggs. Wandering males are clearly attracted to the *chiff* calls of fertile females, and the high-speed chases sometimes do result in successful copulations. From a male's point of view, the calls of the females allow him to quickly locate fertile females that otherwise are well hidden among the dense trees. For the female, her calls are an invitation to wandering males to try their luck in head-to-head competition with her mate: may the best man win.

2   MONOGAMY IN A TROPICAL PARADISE
Timing Is Everything

In 1996, my husband and I were in Panama with my new graduate student, Owen Moore, to start a research project on mangrove swallows. Our small boat came around the corner of Barro Colorado Island, where the Smithsonian Institution operates a tropical research field station, and suddenly we hit the waves. It was mid-morning in late January, and the dry season winds were already blowing fiercely down the length of Gatún Lake. Buckets of water flew over the gunwales with each pounding and soon we were all soaked to the skin, along with our gear. It seemed like ages, but was only twenty minutes, before we arrived at the sheltered south side of the island and pulled up to a dead tree snag.

Gene and I were there helping Owen put up several dozen nesting boxes for mangrove swallows. There wasn't much room in the boat with three people, a stack of nest boxes, and our backpacks. Mangrove swallows nest in tree cavities along tropical lakes and rivers, and Owen would spend the next two years studying their mating habits and breeding biology.

Barro Colorado Island was formed in 1914 when the Chagres River was flooded to form Gatún Lake and the Panama Canal. It feels surreal to bob peacefully in a boat with a lush tropical forest within arm's reach, listening to howler monkeys roar in the distance, only to look up and see a massive container ship passing by on the lake. Most of the dead tree snags at this end of the island are decades old, once-magnificent mahogany trees that were killed by the original flooding.

Gene claims to this day that we very nearly were killed that morning. We had dried off quickly in the relentless sun, and even though I had smeared myself with sunscreen I still felt like my skin was slowly roasting. Owen was our captain and steered the boat toward a promising snag. As we puttered toward the dead tree, he cut the engine, and we heard Gene say, "Don't hit this snag . . . something doesn't look right." The boat hit the tree with a sickening *thud* and a dark mass suddenly poured out of the top about 6 metres above our heads. It reminded me of smoke coming out of an old steam engine, expanding as it gets farther from the funnel. Gene yelled, "Killer bees! Get out of here, fast!"

I had never seen Gene so scared, and he later confessed he continued to have nightmares about this for several weeks. Just a few years earlier, two fishermen on Lake Gatún had been killed by a bee swarm. The story goes that they had jumped into the water to escape the bees, but were stung on the face hundreds of times when they came up to take breaths. As we watched the bees pour out of the dead tree, Owen panicked and pulled frantically on the starting line, flooding the outboard engine. One, two, three, four pulls before the engine roared to life, he threw it into reverse, and we backed away—just before the cloud of bees swirled down to the base of the snag.

After our bee encounter, and finding that wasps liked the nest boxes almost as much as did mangrove swallows, Owen always opened the door to a nest box with a can of wasp and hornet spray aimed and ready. He checked the nest boxes twice a week to find out whether the females nested in a sudden burst, like a typical northern spring. He found that some females started their first nest in January whereas others took their time and laid eggs in March or April; collectively the egg-laying efforts were spread out over a four-month period. He also caught females and males in their boxes to band them and take blood samples for DNA testing.

The reason we were so interested in the behaviour of the mangrove swallow was that their northern relative, the tree swallow, has been studied intensively and thus we had a point of reference. The first bird I ever banded was a tree swallow; I spent four summers as a student checking tree swallow nest boxes, marking adults, and watching the details of their lives. At the time, I could not have imagined that one day I would have my own student doing the same for mangrove swallows, far away in Panama. Tree swallows also nest in tree cavities, so they are aptly named, but most eggs in the population are laid during a two-week period in spring. Extra-pair matings are rampant in tree swallows and

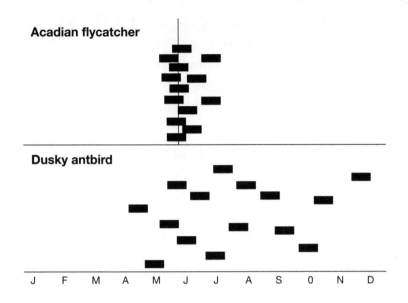

Figure 2.1. *Migratory birds, such as the Acadian flycatcher, have synchronized fertility windows (indicated above by black bars). Most tropical birds, such as the dusky antbird, breed almost year-round.*

about three-quarters of females produce at least one illegitimate young. We expected the tropical mangrove swallow to be largely monogamous because of its long breeding season.

The length of the breeding season is important; when most females in the population are fertile at the same time, competition among males for fertilizations is at its highest. Male birds face the same challenges fertilizing eggs whether the female involved is their own mate or someone else's. Spring in northern regions is heralded by a rush of frantic nest building and egg-laying (Figure 2.1) and brings an onslaught of sexual receptivity that is a recipe for promiscuity. During a single week in late May, for a typical migratory bird like the Acadian flycatcher, a male is likely to be surrounded by fertile, and receptive, females. A secret trip off-territory to find willing partners is likely to pay off because opportunities abound.

In tropical regions, though, most birds nest throughout much of the year and so neighbouring pairs are largely out of sync with each other (Figure 2.1). The dusky antbird, a small tropical forest bird we were also studying in Panama, is typical of many tropical forest birds; fewer than 10 percent of females are fertile at the same time and extra-pair mating is rare. A male dusky antbird would likely receive a cold shoulder even if he bothered to leave his territory and court a neighbouring female, because chances are she would not be getting ready to lay eggs.

We expected that the long breeding season of mangrove swallows, and asynchronous breeding among females, would be matched by relatively low rates of infidelity. It was satisfying to see Owen's lab results, which showed that tree swallow young were three times as likely to be the result of extra-pair mating as are mangrove swallows.

My next tropical student, Sharon Gill, spent three years in Panama, from 1997 to 1999, studying mate choice in the buff-breasted wren, a tropical bird that lives in secondary forest from Panama to southeastern Brazil. These feisty little wrens have a similar lifestyle to the dusky antbird, in that they are paired year-round, breed during the rainy season, and the pair sings, nests, and defends territories together. She worked in a small forest patch along the Chagres River, near the Panama Canal, that is home to some two dozen pairs of wrens. For three breeding seasons, Sharon banded all the adults, found and monitored their nests, watched pairs, took DNA samples, watched, and listened.

In three hundred hours of observations Sharon witnessed only three instances of what might have been an attempt at a sneaky copulation, where a single bird intruded onto a territory. When females were fertile, males did not suddenly start following females, singing more often, or otherwise show signs that paternity was at risk. Pairs usually stayed within a few paces of each other, but this close association was mutual rather than a result of males following females to ensure exclusive mating access. At any given time in the breeding season, only 10 percent of females were fertile; as expected, Sharon found that only 1 of 31 nests contained illegitimate offspring.

Comparisons across species allow us to understand what aspects of a species' ecology or behaviour are tied to the evolution of particular traits, in this case cuckoldry. Once DNA testing became technologically possible and even routine, in the early 1990s, a rush of scientists learned how to take blood samples and began paternity testing en masse. Well over one hundred bird species have been studied using genetic analysis of paternity, showing that most birds feature extra-pair mating. The group of birds with the dubious honour of being most active in cuckoldry is the passerines, or songbirds, where 86 percent of species are guilty of frequent infidelity.

It is almost a relief, then, to learn that true genetic monogamy is the rule in some parts of the bird world. The list of the faithful includes a wide variety of birds like the barnacle goose, chinstrap penguin, common loon, Wilson's storm petrel, New Zealand robin, silvereye, and Carolina wren. Unfortunately for us, though, it seemed that no one else was especially interested in studying the idea that tropical birds are also monogamous, so for the time being we were on our own. So far, we had produced three examples of tropical birds with long breeding seasons and low, or no, cuckoldry. When these species were included in an overall comparison, the rate of infidelity across songbird species clearly increased with the degree of synchronization among nesting females.

One doesn't necessarily need a fancy DNA lab to get an idea of the extent of male sexual competition over females. In most animals, including birds, the size of the testes is a pretty good predictor of how many partners, on average, a male copulates with. In species where females are promiscuous, males copulate frequently and need to be well prepared with a large stockpile of sperm. In strictly monogamous species, the testes are suitably modest in size because copulation occurs rarely. The variation among species can be appreciated in mammals, for instance, by visiting the sheep pens at the county fair; males have enormous testicles that dangle halfway down their legs.

Our first evidence that tropical birds are often genetically monogamous came from measurements of testes, and not the more expensive, and harder-to-get, DNA analyses of family groups. One cannot simply whip out a ruler and measure bird testes, however, because they are hidden inside the bird's body. Flight requires birds to be streamlined and external testes, as we see in most mammals, would increase drag during flight. Most testes measurements come from birds dissected long ago, then stuffed and laid out neatly in museum drawers.

The testes are usually easy to find inside a bird, and look like two shiny, white, bead-like organs that lie against the back of the body cavity. You would not be able to miss the large testes inside a breeding tree swallow, for instance, because the globes occupy a large portion of the gut cavity. Despite a similar body size, tree swallow testes are over fifteen times larger than those of the mangrove swallow, like comparing a garden pea to a pinhead. Tree swallow males are not shy when it comes to mating, and males copulate often with their partners (five to ten times an hour is not unusual), in plain view, and with much chattering and commotion.

The intense competition to fertilize eggs, both of their own mate and of other females in the population, means that a large supply of sperm is needed in promiscuous species. In contrast, we found that most tropical songbirds have tiny testes, even in the middle of the breeding season, because mating is typically restricted to their own mate.

~~~

My first trip to Panama with Gene was to study the dusky antbird, a small bird that lives year-round in forest edges and regenerating forest. For many years, we spent every winter in Panama and took our family with us. Douglas learned how to walk in Panama and many of his first steps intercepted the paths of the leaf-cutter ants. A column of ants frequented the mango tree behind our building and their millions of footsteps had trampled a 10-centimetre-wide path through the grass. Ants heading back toward their underground nest carried small sections of leaf that they had cut from the tree. Douglas, clad only in diapers though it was the middle of February, enjoyed standing on the ant highway and watching them climb over his unsteady feet.

We lived in Gamboa, a tropical oasis where the highway ends and the town is surrounded by the Soberanía National Park, the Chagres River, and the Panama Canal. Gamboa lies within the former Canal Zone and the homes were originally built for the U.S. citizens who lived there to operate the canal. The town is unlike a typical chaotic Latin American village, and has neatly cared-for lawns, and palm trees lining the streets, and used to have a golf course, swimming pool, gymnasium, theatre, tennis courts, and other suburban luxuries. Although now entirely Panamanian, and lacking most of these amenities, Gamboa still retains its Canal Zone feel and is home to many biologists working at the Smithsonian's Tropical Research Institute.

The houses in Panama are designed for coping with the long wet season that lasts from May through December. To escape the sodden ground during endless rains, they are raised up one level so that the living areas are on the second and third floors and the car park is underneath the house. Large screened windows cover most of the outer walls, and a ceiling fan in every room provides much-needed circulation. The roof overhangs the windows by a generous amount so the windows need not be closed during downpours. It is so warm, even at night, that there is no heating system, and one rarely, if ever, closes the windows. Surrounding the house is what looks like a small concrete moat that catches the water pouring off the roof during heavy rains.

Gamboa consists of one hundred or so homes, the Dredging Division, and, at that time, a single tiny store and gas station. Douglas and Sarah used to beg us for a centavo when they heard the rhythmic honking horn of the white van that was laden with bread and sweet rolls. We called the driver the "pan man," and the kids rushed out to the street to wave him down, clutching their coins. With limited offerings in Gamboa, he had plenty of customers and his thirty-minute drive from Panama City was well worth it.

For our part Gene and I got excited about the little red Toyota hatchback that came to town on Sunday afternoons, though the driver was strangely stealthy. The man always stopped in front of our house, popped his trunk, and waited for us to come out to look inside the cooler. He usually had fresh corvina (a mild and delicious snapper), fresh shrimp, and, if we were lucky, *ceviche* (raw corvina, octopus, or other seafood in lime juice).

Our daily routine started with papaya, toast, soursop (guana-bana) yogurt, and a strong cup of Cafe Duran strained one cup at a time. A local woman named Ninfa arrived at 7 a.m. to take over the home front, and Gene and I piled our field gear in the boxy old red Jeep for the short drive to the rainforest. Ninfa knew that we studied "*pajaros*" but otherwise was mystified by the long net poles and parabolic microphones. It's fair to say she was downright disgusted with the box of squirming *gusanos,* our mealworms, which lived in a box on top of the refrigerator.

We looked an odd sight, too, especially me, as most Panamanian women make a point of dressing well. In the field I usually wore white pants, black army boots, and had masking tape wrapped around my ankles with the sticky side out. The pants and tape were a defence against ticks. At least once a week I would look down to see dozens of tiny dots on the tape, with many more marching northward to my vulnerable waistline and T-shirt. It was a race against time to use extra pieces of masking tape to press onto my pants, trapping the rest of the horde. Search for escapees was a daily grooming ritual.

The boots were a defence against various biting creatures, including snakes, ants—and my worst enemy—tiny invisible larval mites called chiggers. Chiggers are overwhelmingly abun-dant in grassy areas and the dry leaf litter of forests, and the microscopic critters harmlessly chew on skin. The only sign that chiggers were ever present are the dysfunctional itchy red welts that form soon afterwards. Chiggers crawl through the tiny holes in socks and pants and tend to pile up and take a bite where the clothing is tightest. After a few weeks in Panama my ankles, waist, and bra line were adorned with hundreds of itchy chigger bites, which became truly miserable on hot humid nights.

Because dusky antbirds prefer forest edges and young forests, we spent most of our time working at the beginning of, and along the Pipeline Road, a dirt road that penetrates far into the Soberanía National Park and was built to establish a safe oil supply during the Second World War. Many of the dusky antbirds lived in forest patches that lay between the navigation cuts, which are long, wide corridors of grass that stretch from the hilltops down to the canal. Lined up along the corridor are a series of large, white diamond-shaped signs with black lines and crosses— the ships' pilots use these to aid in navigating the narrow passages through the canal. We had to show our Smithsonian U.S. Government ID cards at the guardhouse before being allowed into the navigation cuts. Some of our birds lived near the old concrete ammunition bunkers where Pipeline Road began, and the rest along the edges of the dirt road. The deeper one goes into the forest, the narrower, bumpier, and less passable the road and the more precarious the bridges over the streams. The road is travelled by biologists, ecotourists, and, on weekends, a handful of mountain bikers.

Dusky antbirds are small, dull-coloured birds that skulk in the thick undergrowth and are more often heard than seen. Males are dark grey and females are grey-brown, but both have a hidden ornament used to threaten rivals. When angry, a dusky antbird fluffs up its back feathers and hunches over, revealing a bright white back spot. Its sturdy beak is used to capture, and shred, large insects.

Many species of tropical birds have *ant* in their name: antbirds, antwrens, ant-thrush, ant-vireos, ant-shrikes, ant-tanagers—so much so, you might be tempted to think there must be few ants left in Panama. These birds do not eat ants, however; they are so named because some species in this group follow army ant swarms. Army ants form colonies of millions of individuals, and the insects that are usually well hidden run for their lives when an ant swarm sweeps across the forest floor. The birds perch in the low branches above the ants and feast on these insect refugees. Dusky antbirds are not professional ant followers, and instead defend modest-sized territories to ensure a steady supply of insect food over the course of the year.

Dusky antbirds, like most insect eaters, breed during the rainy season (May to December) when food is plentiful. A pair lives on its territory year-round, and the male and female do almost everything together: sing, forage, build a nest, incubate eggs, feed young, sleep, and fight off intruders. The bag-like hanging nest is built somewhere among a tangle of vines, and the female lays only two eggs, a clutch size typical of many tropical passerines. The long breeding season means there is no particular hurry to build a nest as soon as the rains start. About 90 percent of nests are eaten by predators so most pairs must re-nest many times during a season, and even then often produce no young. The offspring, one or two at most, live with the parents throughout the rest of the breeding season and the following dry season. Our DNA testing of the handful of successful nests showed that dusky antbirds were genetically monogamous.

One of our study sites was along Old Gamboa Road, which runs parallel to the canal through a beautiful old-growth rainforest. One morning we had been doing our annual census of dusky antbirds, to look for banded birds, and came across a large mixed-species flock of birds feeding at a fruiting *Miconia* tree. Golden-masked tanagers, plain tanagers, and red-legged honeycreepers were helping themselves to the all-you-can-eat buffet. A troop of Geoffroy's marmosets, colourful little primates the size of a house cat, came tearing by, barely pausing long enough to scold us.

Less than an hour later, we were treated to another mixed-species flock, this time an insectivorous one. Tropical gnatcatchers, forest elaenias, and white-winged tanagers were sticking close to the lesser greenlet, a small plain-looking bird who on first glance makes an unlikely flock leader. Up in the canopy above the flock was a troop of white-faced monkeys, crashing from branch to branch. Perched nearby, I got my first-ever look at a double-toothed kite. This raptor is somewhat like the professional antbirds, but it follows monkey troops, not ants, and catches the insects and lizards that are stirred up by their movements.

Two weeks later, when we returned to that exact place on the Old Gamboa Road, we found a bulldozer and a cleared forest. The canal was being dredged and widened in the bottleneck called the Gaillard Cut, and the forest was being cleared to make a convenient dumping ground for the dredgings. We never found out what happened to the dusky antbirds who had lived in this forest their whole life; our banded birds were never seen again, even though we searched the remaining forest down the road.

Our tropical research on mangrove swallows, dusky antbirds, and buff-breasted wrens all supported the idea that prolonged breeding seasons tend to promote monogamy. For a different kind of test, Gene and I decided to study the clay-coloured robin in Panama because they do not defend territories or remain paired year-round, males do not sing outside of the breeding season, and breeding is quite synchronized. The clay-coloured robin has an abrupt onset of breeding rather reminiscent of spring in temperate regions, so we were expecting to find that this tropical bird was an oddball and featured a high rate of extra-pair mating. It was probably no coincidence that in our earlier comparison of testes size, the clay-coloured robin had stood out as exceptionally well endowed among tropical birds. The real proof of infidelity, however, could come only from suburban fieldwork to collect blood samples for DNA testing.

There is not much glamour to be found in studying clay-coloured robins, for they nest in parks and backyards in towns. This means that instead of spending one's time roaring around in boats or hiking through a rainforest, the field biologist must tolerate a day of sidewalks, stray dogs, and muffler-challenged buses. Even the bird's appearance is dull, a uniform chalky brown from head to tail, and it is so common that it is typically overlooked.

These robins are known locally as "rain birds" because it is said they call for the rains to begin. During the height of the parched dry season in February, over a period of a few days, all the males begin singing in earnest in the pre-dawn darkness, a chorus that marks the beginning of breeding. Silence one week turns into a full song the next week, followed soon after by females working frantically at nest building. Though the robins are present year-round, they pick the driest time of year to begin breeding. Adult robins can survive on a diet of fruit, which is plentiful during the dry season, and begin breeding well before the rainy season so as to avoid the high levels of nest predation that occur then.

Nest searching involved wandering around town looking for a robin on the ground gathering a huge mouthful of grass. It was usually easy to keep her in sight as she flew across a lawn to a nearby fig tree, cashew tree, or eave on a church or house. Catching robins, however, proved to be a little more difficult than I had imagined. Most nests were too high to simply place a mist net nearby, and even then our mist nets were near useless in open areas where the wind and sun made them look impenetrable rather than invisible. My only recourse was to use nest traps; if you can convince the bird to enter the trap, it inadvertently steps on a foot pedal and the door drops behind it.

During the dry season, robin parents will do almost anything to get a juicy worm for their chicks. After a month or more of sunny days, the ground becomes so parched that large cracks form in the earth and underfoot it feels like concrete. The creatures of the soil disappear far underground, out of reach of robins. Although adult robins are perfectly healthy and happy eating fruit, they must feed insects and worms to their newly hatched nestlings because they need the protein.

The main Smithsonian research building in town, where we got our daily fix of e-mail, was home to a popular bird feeder, stocked à la tropics with banana and papaya rather than birdseed. The feeder attracted a steady stream of tanagers, honeyeaters, motmots, and woodpeckers, and many of our robins. One day I put a handful of mealworms on the feeder and discovered that robins quickly become addicts. I saw one robin with a beak full of mealworms fly off down the hill behind the building; minutes later it was back for a refill, a sure sign it was feeding nestlings. Over the next hour I followed the bird farther each trip, because it took the same route each time. Her nest was four blocks away, on the side of a house at the edge of town.

This was the key to catching adult robins—bait the traps with commercially available mealworms. It was agonizing to watch a bird hop around the trap, and on top of it, tilting its head to look at the mealworms with one eye but not managing to figure out the location of the front door. One of my nests was in a mahogany tree beside the Panama Canal Commission's fire station. I had caught the male within an hour, but for three mornings I grew more and more irritated as I watched the female dance around my trap, picking out the occasional escapee, before she finally wandered into it. All the while, on the other side of the fence, the firemen were playing cards, washing the trucks, and glancing curiously at the strange woman sitting on a stool under the tree.

After I had the blood sample from the female, it was time to sample her brood of nestlings. The catch was that the nest was 20 metres high in the huge mahogany tree under which I had been sitting. As Gene and I discussed our predicament, the fire chief came to the fence and asked what we were doing. He listened carefully as Gene explained what the traps were for and then smiled as I pointed to the nest in the tree. "I think I may be able to help you," he said. "Sundays are pretty slow and my men could use some extra training."

Ten minutes later a gleaming red fire truck backed up to the tree, and two men manoeuvred a long extension ladder that was on top of the truck, leaned it against the tree, and one of them climbed the ladder to the nest to retrieve the nestlings for us. Ten minutes later the nestlings were back in the nest, and the firemen went back to their Sunday morning routine.

A year later, after the lab testing was finished, I learned that this nest was among the 50 percent of nests that contained illegitimate young.

~~~

The idea that breeding synchrony can explain why some species thrive on cuckoldry while others are typically monogamous is somewhat controversial in ornithological circles. Since Gene and I first published the idea, along with evidence showing small testes size in tropical birds, dozens of review articles have criticized the idea and studies have been published on various bird species showing little or no effect of breeding synchrony. One not-so-anonymous reviewer of my article on clay-coloured robins suggested that my prediction that a synchronously breeding tropical bird will have extra-pair fertilizations qualified me for membership in the Flat Earth Society. Part of the controversy is probably our fault, because in our original 1995 article Gene and I deliberately went out on a limb and claimed, "We believe breeding synchrony is the most important factor promoting the evolution of extra-pair mating systems."

Comparisons of cuckoldry rates among songbird species do show a strong correlation with breeding synchrony, and several studies have shown that within a population the synchronously breeding females are more likely to produce extra-pair young. This is a great example of how a complex behaviour cannot be explained by one simple factor. Breeding synchrony is but one of many behavioural and ecological factors that influence whether or not a female will benefit from infidelity, whether she can get her own way, and whether males benefit from spending time and effort looking for extra-pair mates.

I recently had a graduate student, Melissa Evans, who set out to study the mating system of the wood thrush, a songbird that breeds in our Pennsylvania forest. Wood thrush are renowned for their rich, complex song and our original idea was for Melissa to relate female mate choice and trips off-territory to the quality and quantity of male song. Were the successful males those who sang earlier at dawn, more often, or had more complex songs? Our earlier studies had shown that many of the other songbirds in our forest, like hooded warblers, Acadian flycatchers, red-eyed vireos, and American redstarts, had high rates of cuckoldry, so we were confident that the wood thrush would be an excellent bird for her thesis research.

Melissa mounted radio-transmitters on both the male and female of a pair, so she could track them simultaneously. Do females wait for their own mate to be absent before sneaking off themselves? Are females more likely to visit particular neighbours with the best voices, and also get fertilizations from those males? Melissa found plenty of extra-pair behaviour during her first summer of fieldwork, with both males and females leaving their territories regularly. Imagine her surprise, though, when her lab work that winter revealed that only 6 percent of the offspring were the result of cuckoldry.

Despite breeding synchronously, wood thrush do not have abundant extra-pair matings because, as Melissa discovered, males follow their mates on and off the territory. Tracking data showed that females left their territories only when fertile, as expected, and visited neighbouring males almost once per hour. However, for most of these female trips, the mate was following closely behind. A male spent at least 90 percent of his time within 10 metres of his fertile mate, close enough to detect intruders and prevent extra-pair copulations. Mate guarding can sometimes trump the influence of breeding synchrony, blurring the patterns we look for among species.

It is this complexity of behaviour that makes it so interesting to study birds. Just when you think you have them figured out, something unexpected happens and the questions start all over.

3 FINICKY FEMALES
What Makes Males Look Attractive

Early mornings are par for the course for a bird detective, and on this morning I was driving in the darkness from my farmhouse in Pennsylvania to the nearby town of Edinboro. I arrived at a small trailer park by the lake, and walked down the long driveway with gravel crunching loudly underfoot. I whispered greetings to my students and took a place at the card table. There were eight people in our group and we were bundled up in warm jackets even though it was early July. Some of us wore headlamps and one of the cars had been parked facing us with headlights left on. The table was set neatly with clipboards, rulers, calipers, syringe needles, and microscope slides. The centre piece was a wooden block with two long stiff wires arching upwards, a cork impaled on each one. A few stars twinkled above, and the faint glow on the horizon hinted at the burst of activity that would soon begin.

We had gathered for our sixth "martin morning" of the summer. Patrick Kramer, my PhD student, walked slowly toward us straining at the weight of the large suitcase-like wooden box he was carrying. This "hotel," as we call it, is made up of twenty-four separate compartments, each with a hinged flap door at one end and a screen window at the other. Inside the hotel rooms were the purple martins that Pat had just removed from their apartment house by the lake.

Purple martins are colonial swallows that readily use man-made housing and have the convenient habit of sleeping inside their nesting cavities at night. We catch the entire colony using a puppet-style nest-trap system that is installed in the afternoon. A small plate is hinged above each entrance hole, and a piece of fishing line runs up through an eyelet to hold the door open. All twenty or so trap-door lines are fed to a single point, the doors are raised, and the line is tied on the pole below the house.

The martins had gone to bed at sunset with the usual commotion and grating alarm calls of *chee, chur-chur-chur, chee-chur* as the birds swirled around the houses before ducking inside. At midnight, Pat had tiptoed up to the house and cut the line, listening for the satisfying soft *clunk* as all twenty doors dropped simultaneously. The birds, unaware of our trap, slept on through the night. Before dawn, the house had been lowered and the doors to each compartment had been opened one by one to transfer the puzzled occupants to their own hotel room to await processing.

Pat sat at the head of the table, while the rest of us formed an assembly line on either side. Each guest first received a permanent identification number in the form of a numbered metal bracelet around each ankle. Pat pulled the corks off the stiff wire posts and slid a nine-digit aluminum leg band from one post and a purple band with a matching number from the other post. After a couple of gentle squeezes from the pliers, our first bird, an older male with steel blue plumage, would forever be known by the awkward nickname of "purple left C288."

Once banded, the martin was passed down one of the assembly lines. The first person weighed the bird on a small scale, and carefully measured the size of its body with a ruler. John Tautin, director of the Purple Martin Conservation Association, had the critical job of manning the clipboard and entering data as statistics were barked out: "C288 male: 50.2 grams, wing 145 millimetres, tail 67, tarsus 17.3." The next person, after cleaning the outstretched wing with an alcohol swab, took a small blood sample from a vein. The last person in line transferred the blood to a vial, numbered it with the bird's band number, and then prepared a microscope slide with a thin veneer of blood. A runner waiting at the end of the table returned each bird to its hotel room.

By 6 a.m. the headlamps were turned off, by 7:30 the jackets came off, and by 8:30 the martins had all been returned to their nesting house and were free to come and go as they pleased. They settled down quickly to their usual morning routine of commuting out over the lake to find insects to feed to their hungry nestlings. Our unwelcome and brief interruption of the bird's breakfast was necessary to gather a wealth of clues and evidence that would help us solve many mysteries about martin behaviour.

From Texas to Florida and north well into Canada, hundreds of thousands of backyards are proud home to purple martin houses. It is easy see why purple martins are front-runners for North America's favourite bird. They are graceful and acrobatic flyers who swoop into their entrance holes at full speed, closing their wings at the last possible moment. High in the air they twist, turn, and dive as efficient and ruthless predators of unsuspecting butterflies, dragonflies, and other aerial insects. The iridescent steel blue colour of older males is as pleasing to the eye as their rich, gurgling song is to the ear. But it is really the social behaviour of the martins that makes them win hands-down. At a martin colony the action never stops.

An observer might see a male and female sitting side by side inside a cavity, both largely hidden from view. This pair bred in this colony the year before, though not with each other. A young male has just landed on the porch two doors down. This youngster has female-like plumage, a brown back and white belly, but also a spattering of telltale blue feathers. The young male keeps his feathers sleeked down and wings tight against his side, as he ever so slowly peers into a vacant compartment. In a flash, the older male wriggles out and charges down the porch to confront the cheeky intruder, who wisely flees.

On the ground below, a female lands near the lake and pokes at a thin stick. Her mate, a young male, stands guard beside her, watching closely as she balances the nesting material in her beak. As she flies back up to the house, he follows behind. She begins to enter the house but is stopped because the stick is too long to fit through the hole. Time and again she thrusts forward, only to have the stick block her entrance. The comical scene ends by accident when she loses her grip and finds herself holding the end of the stick, which fortuitously swings alongside her as she goes inside. A few minutes later she returns to the lake for another load, tailed again by her wary young partner on the lookout for older males who are known to ambush females to sneak a quick chance at mating.

At the house next door, a wing tip momentarily sticks out of a hole only to disappear. Suddenly an entire head appears as a young male desperately tries to escape the nest cavity. He cannot get out because he is in the fierce grip of another male, who manages to pull the frantic bird back inside. In the shadows, the challenger takes a beating as he is pecked repeatedly on the head and back. The battle continues for a few more minutes until the intruder gets one wing and his head outside and, finally, after a long struggle, breaks free. Afterwards the owner sits calmly at his entrance hole, only his shiny, steel blue head protruding, as if daring the intruder to return.

To the casual observer, the antics of purple martins provide endless entertainment and fascination, not to mention concern for the tenants. Questions immediately come to mind: Why do young males look like females? Why do older males hog extra nesting cavities? My students and I have our own seemingly simple question about purple martins: Why do females prefer to mate with older males? Earlier studies by another purple martin fanatic, my husband Gene, showed that females socially paired to an older male in adult plumage are genetically faithful, whereas those paired to a young male are not and have older males as extra-pair mates.

One theory is that older males use their steel blue iridescent plumage to advertise their age and hence their ability to survive at least two years. Survival is no easy feat, since in migratory songbirds about half of the population dies each year as a result of the long, arduous journey. Parasite infections are one of the leading causes of death in young birds, and so any bird who survives to its second breeding season likely has a good immune system as a result of the genetic instructions inherited from its parents.

Our banding operation, combined with watching birds at

their nest sites, allows us to identify the dozens of couples who live in the colony. We check the nests twice a week so we know how many eggs each female laid, how many hatched, and how many nestlings successfully left the nest, each sporting its own leg band. In the past fifteen years the Purple Martin Conservation Association has banded about 16,000 nestlings in northern Pennsylvania. Each spring, Pat and his helpers scan dozens of martin colonies with a telescope, looking for banded birds to find out which nestlings sampled the year before have survived their first year of life. Does it matter who your daddy was? To find out, we will compare the long-term health and survival of martins sired by older, proven males with that of nestlings whose genetic father was a young, first-time breeder.

～

Females are picky about which males they pair and copulate with, and over many generations female choice has driven males in an escalating race to outperform one another. This process, called sexual selection, is a potent force in evolution and is responsible for many of the bright colours, spectacular songs, and ornaments that make male birds so attractive to human observers. Females are faced with a bewildering array of male characteristics, each of which may indicate a different underlying quality. Males are usually at the mercy of the female and must compete with other males for female attention.

The preparations for courtship can be elaborate. Imagine that a gentleman has met an attractive member of the fairer sex and has invited her over for an intimate dinner. The bachelor apartment is tidied up and vacuumed, the clutter hidden in closets and drawers. He combs his hair just so, and dresses carefully in slacks and a pressed shirt. A small bouquet of flowers is placed casually on the dinner table and soft music sets the mood. But what if she is not impressed? If he were to take a lesson from the bird of paradise he would slip on his black tutu and prance around the room. If he were imitating a bellbird he would sidle up to the cautious female, wiggle his dangling worm-like cheek ornaments, and then yell in her ear; if a frigatebird he'd inflate a gigantic red balloon, attach it to his throat, then jiggle it around enticingly while jabbering loudly.

The basic principle of mate choice theory is that females prefer male traits that are difficult for males to grow or display. Why? By paying attention only to elaborate dances, plumes, and songs, females can accurately test male quality. Males who are weak, or sick, will not be able to perform energetically costly displays like twirling, jumping, and leaping as well as will healthy males. One of the most prevalent forms of illness in birds comes from parasites like avian flu and malaria, and various gut parasites. If a male's displays are handicapped by parasitic infections, the female could judge his current health and, less directly, his genetic resistance to disease.

The blue-black grassquit is a small, monogamous songbird with a conspicuous male courtship display. Males defend small territories in tropical grasslands, where pairs nest close together. The male has an iridescent blue-black plumage during the breeding season, and performs a wild courtship display that consists of repeatedly leaping upward from a perch and flashing the white patches under his wings. Males vary greatly in the height and frequency of their leaps, but is this a cue females could use for successful mate choice?

Researchers from the University of Brazil caught males and females during the non-breeding season and moved them to an aviary. The wild-caught birds had feather lice and a common intestinal parasite called coccidia, microscopic organisms that in serious cases cause bleeding and diarrhea. Half of the grassquits were randomly chosen to receive a suite of medicines to treat these parasitic infections, and the other infected birds were left untreated. During the breeding season, an infected and a treated male were placed in separate compartments in an aviary, and allowed to see a female in an adjacent cage through a glass partition. Males started displaying spontaneously, even though the setting was far from natural. Treated males, who had far fewer parasites, spent more time displaying, jumped higher, and jumped more often than infected males. A sick male cannot conceal his condition during courtship, and the display is a reliable signal for females to use in mate choice.

For female birds, "seeing red" can be a good thing when choosing a partner. The bright yellows, oranges, and reds we see on bird beaks and feathers are made with carotenoid pigments. Birds cannot manufacture carotenoid pigments and can get them only through their diet. The intensity of red and orange colours in bird feathers and body parts (for example, beaks, wattles, feet) are an honest signal of a male's ability to find and consume foods rich in carotenoids. In captive studies, or pet birds, those deprived of carotenoids in their food soon fade in colour and become noticeably paler. Most body parts can continuously acquire and release carotenoid pigments and so are sensitive to changes in a male's diet and health. Feathers, on the other hand, are inert once they grow, and reflect the male's access to food at the time those feathers developed. Since birds moult their feathers only once or twice a year, and often long before the breeding season, feather carotenoids represent a male's historical rather than current condition.

Eating carotenoid-rich foods does not guarantee a male will have sexy colours, because carotenoids are also used in the immune system. Carotenoids have potent antioxidant properties and help to protect cells and tissues from damage. Carotenoids also directly stimulate the immune system of animals; carrots *are* good for you. Only a healthy male with plenty of carotenoids can afford the luxury of storing them in his feathers and body parts. Thus a female who bases her mate choice decisions on carotenoid colours would gain by pairing with a male in good condition, and possibly one who also has genetic resistance to diseases and parasites.

Geoffrey Hill, a professor at Auburn University in Alabama, has done pioneering research linking carotenoids to male colouration, mate choice, and disease resistance. He studies a common urban songbird, the house finch, which he calls "a red bird in a brown bag." The male house finch has a brown streaked back but an ornamented head, throat, and rump that can range from yellow to bright red among males in the same population. Mate choice experiments in the field and lab have shown clearly that females prefer bright red males whether they are natural redheads or are reddened with hair dye. Males who are denied a carotenoid-rich diet moult into a pale yellow colour and subsequently are less preferred as mates.

In an aviary experiment, male house finches that were experimentally infected with a bacterial pathogen called *Mycoplasma gallicepticum* grew feathers that were more yellow, and less red, than those of uninfected males. Though this shows that males who are sick cannot grow red feathers, it remains unclear to what extent carotenoids tell a female whether a male has low genetic resistance to disease (which could be passed on to her offspring) or simply that the male has a temporary illness.

Mycoplasma gallicepticum is a real-life, and recent, problem for house finches. The bacterium infects poultry, and was first reported in a wild songbird in 1994 when people began seeing diseased house finches at bird feeders in Maryland. The bacterium infects the respiratory system, and has serious health effects, but is easily visible because it causes grotesque swellings due to eye infection. The disease quickly spread throughout the eastern population of house finches, tracked by backyard birdwatchers who reported their observations through the citizen science project FeederWatch, run by the Cornell Lab of Ornithology.

To find out whether red colouration predicts a male's ability to fight a novel disease, Hill captured house finches in Hawaii, where *Mycoplasma gallicepticum* is not yet present, and brought them back to the lab in Alabama. After three weeks of getting used to their indoor home, half the disease-free birds were deliberately infected with the bacterium. Birds were observed three times a week to score the severity of eye infection. All males developed eye infections, regardless of colour, but after six to eight weeks the redder males were able to clear the infection faster than yellow males. This is one of the few studies showing red colour in a male is not just a litmus test of his current health but also tells a female of his, and thus her future offspring's, ability to fight disease.

Careful mate choice can also benefit a female directly via the resources the male has to offer to her, rather than indirectly via his DNA. This direct benefit applies mostly to monogamous species, where the male plays a critical role in raising the family. In ducks and geese, where the young leave the nest upon hatching, his role may be of bodyguard. In songbirds, where newly hatched young are naked and helpless, the male helps to feed the nestlings and fend off predators. These kinds of benefits too increase the health and survival of the offspring.

Females can also judge a good provider by his colours. Differences in yellow, orange, and red colouration among males tell a female how much access the male has had to carotenoid-rich foods, either because he controls a good territory and/or because he is efficient at foraging. For females who require extensive male help in feeding young, choosing males based on their parenting skills may be especially important. The difficulty, though, is that female choice precedes a male's care of nestlings by at least several weeks. A female must judge a male's likely parental quality indirectly via his body condition, size, age, and other traits that can be assessed at the time of pairing. Larger males, or older males, are likely to hold higher quality territories and are experienced breeders.

The European blackbird is a thrush common in parks, backyards, and urban areas. Males are entirely black, but bill colour ranges from pale yellow to orange as a result of carotenoid pigments. Researchers in Dijon, France, caught males to measure their bill colour and then observed subsequent parental care and nesting success. Males with orange bills fed nestlings at a relatively high rate, and their mates appeared to match this effort by also exhibiting a high feeding rate. Nestlings in these nests were in good condition, and orange-billed males fledged more offspring than yellow-billed males.

~~~

The common yellowthroat is a card-carrying bully who boldly displays his aggressive colours while challenging rivals with his ringing *witchity-witchity-witchity* song. Though named for the bright yellow throat and underparts, this little warbler is better distinguished by his Zorro-style black mask. Conspicuous and contrasting patches of colours on birds, like the mask, are referred to as badges, or "status signals," and are usually more pronounced on males as a result of sexual selection. Black bibs, crowns, or masks are common in chickadees, warblers, sparrows, grosbeaks, and other songbirds.

The black colour in a feather comes from melanin, a pigment that birds can manufacture themselves in abundance. Why, then, doesn't a male "cheat" and simply make a large badge regardless of his true quality? One theory is that black badges are used mainly in fighting with other males. In that case, only larger or more dominant males can afford to wear large badges; weaker birds would be revealed through face-to-face confrontation and be attacked when they cannot back their macho feathers. If black badges are reliable for male–male competition, then females could also use the badge to assess male quality.

Scott Tarof, now my postdoctoral student, did his PhD at the University of Wisconsin on common yellowthroats. He captured two males with different mask sizes, but who did not hold neighbouring territories, and brought them back to live, temporarily, in the aviary. After one hour of rest, he placed the males together in a room and he observed the level of pushing, chasing, and fighting, and kept close track of who started each fight and who won. Males with larger masks started, and won, more aggressive interactions even though they were not necessarily larger in physical size.

The next experiment was to see which male was preferred by a female, who had also been caught that day and "volunteered" for the task. The males were moved to separate compartments in the aviary, with a small empty room separating them so they could not keep fighting or see each other. The female, in her own section of the aviary, could move back and forth to check out both males. In all ten trials, females spent more time (85 percent) near the male with the large mask.

Since a female could be reacting to the male's behaviour, rather than his mask, Scott then enlarged or reduced mask size to see if this was, in fact, the cue to which females were paying attention. A black felt-tip marker and yellow acrylic paint was used to adjust mask size, and, interestingly, the males did not seem to notice their makeover. Masks were increased, or decreased, in size by about 25 percent. Females were consistent in preferring the males with enlarged masks, even when these were artificially created.

One might reasonably wonder whether the behaviour of a recently caged bird reflects its choices in real life. In the sedge marshes and swamps near the University of Wisconsin, male yellowthroats with larger black masks were usually the first to claim territories. These males were also more likely to attract mates, and were preferred as extra-pair mates by neighbouring females. The reliability of these aggressive signals is put on trial in daily contests and cannot be faked without paying a high price. Females can choose high-quality mates by watching for the visual signals that males use in their battles.

~~~

The beautiful feathers of birds are more than meets the eye, at least as far as humans can see. Humans have three main cone types in their eyes (red, green, and blue) and we perceive a seemingly endless number of hues ("colours") due to the mixing of red, green, and blue light. If we think of this as three-dimensional vision, then birds have four-dimensional vision, because they have a fourth cone type that is sensitive to ultraviolet (UV) wavelengths, which are bluer than blue. This does not mean simply that birds see an extra colour outside our visual range; birds also perceive different colours even within the human-visible spectrum because of the combination of UV sensitivity with red, green, and blue colours.

Many colours that look ordinary to us—say, a white stripe on the head or a shiny black throat—may be afire in UV reflectance. Two birds that look almost identical to the human eye may differ dramatically in their UV reflectance. Ornithologists now use a device called a spectrophotometer to measure the actual reflectance of feathers rather than using the old-fashioned method of matching feathers to paint-colour cards. One can generate a graph that shows which wavelengths are reflected by a feather, but even then humans can never truly appreciate what the feather looks like to another bird. We are like a colour-blind person who is asked to describe a rainbow.

The discovery that birds see one another differently than we see them soon led to the discovery that UV signals are important in mate choice. What exactly does the UV reflectance tell a female about the male? To answer this, we need to understand how the UV-blue colour is formed. Many parents have stumbled over the innocent question "Why is the sky blue?" The blue colour is caused by the physical scattering of light when it comes through the earth's atmosphere and hits air molecules, ice crystals, water droplets, and dust. Light made of short wavelengths, like blue, is most likely to scatter and reflect to the ground, and that is why when we look up we see blue, and not the long red wavelength.

Bird feathers work in a similar way. The UV colours in a feather are not made with a pigment, like melanin or carotenoid, but are a result of the scattering of light inside the feather. There is a spongy layer inside the parallel strands, or "barbs," that makes up the flat surface of feathers. Inside the spongy layer are rods of hard keratin, the same stuff of which claws are made, and air bubbles that scatter the light and produce UV reflectance. Tiny differences in the microscopic arrangement of rods and air bubbles inside the feather change both the UV reflectance of the feather and its brightness.

Male eastern bluebirds are brilliant blue on their heads, backs, rumps, wings, and tails and have a deep reddish brown wash on their breasts. Females have the same colour pattern as males but are drabber. Birds in poor condition apparently cannot grow feathers with the precise structure to generate a brilliant ornament. If parents are forced to raise a larger than usual number of offspring, and therefore exhaust themselves searching for and delivering food, then these birds moult into a drab plumage after the breeding season. On the other hand, birds that are food-supplemented by a curious researcher subsequently moult into a brighter and more ornamented plumage.

A male bluebird that has a more colourful UV-blue plumage is the first to attract a mate, feeds his incubating female and their chicks at a higher rate, and produces more offspring. The most ornamented males are also more dominant over other males in competition for nest sites in tree cavities and nest boxes.

Blue-footed boobies are colonial seabirds who, as the name clearly implies, have bright aqua-blue feet. The name *booby* refers to their ridiculous mating displays, their awkwardness on land, and their very tame nature, which historically often landed them on the dinner plates of early explorers. Males vary greatly in foot colour and show off their feet during courtship displays. During the "salute landing" a male lands in front of his mate and quickly flings up a foot, so he's standing on one leg and holding the other high in the air. From the female's vantage point the conspicuous foot contrasts with the male's white breast. The male also holds a one-bird parade, consisting of exaggerated foot-raising, flaunting the webs upwards and outwards as he marches in front of the female.

Researchers at the University of Mexico captured boobies on an island in western Mexico and masked the brilliant blue feet of some males using a duller blue makeup. Under natural circumstances, male foot colour changes with the male's nutritional state. Females paired to experimental males (those with dull feet) courted less and were 50 percent less likely to copulate than females in the control group. Experimental males were apparently unaware of their non-sexy feet and continued displaying vigorously, so the lack of interest by females was most likely due to the change in foot colour. Thus foot colour is a reliable signal of male body condition.

If male foot colour is a good predictor of his parental care, then bright males should feed their young more and produce young in better condition. This idea is hard to test through simple observation, because it is also possible that the offspring of bright males are superior due to the genetic quality of their fathers (for example, disease resistance). To find out what role genetics versus feeding ability plays in producing high-quality offspring, researchers switched nestlings between fathers with bright blue versus dull blue feet. The cross-fostered nestlings who found themselves with a bright-footed father were in better condition, despite the fact that their genetic father had dull feet.

The signalling system of most birds is finely tuned to the habitat in which they live. Changes in habitat, as result of deforestation, for instance, can lead to a mismatch between a bird's current environment and the historical advantages of specific displays and colours.

The little greenbul is a small, plain-looking bird that breeds in African rainforest. Thomas Smith, a professor at the University of California, compared greenbuls living in pristine mature rainforest versus secondary forest, mostly coffee and cacao plantations, in Cameroon and Equatorial Guinea. Smith found that this shift in habitat is linked to differences in male colour and song, which in turn influences male–male competition and mate choice.

In plantations, males had significantly shorter wings, which could be an advantage in manoeuvring in the thick understory typical of these artificial forests. Birds from plantations also had lower ultraviolet reflectance—enough for birds to notice—of their feathers compared to mature forest birds. When the upper canopy of the forest is removed, more light reaches the lower layers without being filtered through leaves. The light environment becomes richer in short wavelengths and this removes many of the advantages of signalling with UV reflectance. Male song characteristics also were different in plantations, where males sped up their song by delivering more notes per song. Species that occupy more open habitat, where air turbulence can disrupt sound waves, tend to sing at a high rate with rapid sequences of notes. These behavioural and physical changes that can be measured by the field biologist are matched by hidden genetic changes measured in the lab; greenbuls living in plantations are becoming distinct from those that live in tropical forest.

The example of the greenbul shows us how male appearance evolves in the short term to have the optimal, most effective signals for communication. We can expect the same changes are happening in many other forest birds that have been forced into different homes, though they are not currently being studied. Tropical deforestation is so extensive worldwide, and continuing at such an alarming pace, that in a hundred years there may no longer be deep-forest versions of these species.

4 AVIAN OPERAS
Mate Choice by Ear

A few years ago, in April, I found myself in Fargo, North Dakota, as the guest of honour. I had been invited by North Dakota State University to give a special public lecture on why songbirds were disappearing, and was thrilled to see my name on the marquee of the old-fashioned theatre on Main Street. When I checked in with the Dean's Office soon after my arrival, a perturbed secretary asked, very apprehensively, whether I would be interested in seeing something called "prairie chickens." She explained with a bit of a frown that this would involve leaving the hotel at 4 a.m., driving forty-five minutes across the border into Minnesota, and sitting in the cold until sunrise—all this, on the same day that I was lined up for several media interviews, a guest lecture to an undergraduate class and my evening performance. "Of course," I replied without hesitation to the ever more puzzled woman.

The next morning, at 5 a.m., I was walking in the darkness along a well-worn path, listening to the aerial whinnying of a snipe overhead and the two-note whistle of a distant pheasant. A brilliant full moon was low on the horizon, and the flat landscape and trail were bathed in a ghostly pale glow. My guide was Jim Grier, naturalist and retired professor from NDSU, who had spent his career climbing up 60-metre makeshift ladders to observe and check bald eagle nests. Our destination was a small shack about 500 metres away, where we would sit and wait for the arrival of the greater prairie chickens. Jim called this the "Hilton" blind because it had four viewing windows and a padded bench to sit on, a far cry from his rickety platforms high in the treetops. Our blind was in a little bluestem prairie owned by The Nature Conservancy, a tiny remnant of what once covered half of Minnesota.

The prairie chicken used to be abundant on the prairies, but with 95 percent of the native prairie now lost to the plow, the echoing booming calls that signalled spring for the pioneers are now absent throughout most of this bird's former range. Greater prairie chickens have declined by a stunning 91 percent in the past four decades, since systematic bird counts began in the mid-1960s. With the native prairie mostly gone, prairie chickens, like most grassland birds, have to make do with using retired and active pastures and hayfields. Prairie chickens only use sites where the vegetation is at least a few years old, so most fields that are harvested each year are out of bounds. The recent surge in the biofuel industry means that many of the fallowed fields have been put back into production, making it even more difficult for prairie chickens to gather each spring.

As we waited in the blind in darkness, Jim told me tales of engine failure as he was flying over the boreal forest of Ontario surveying for eagle nests. I stamped my feet and rubbed my hands together to try to stay warm, but the borrowed winter coat and boots (from the dean's understanding wife) just barely kept the chill out of my bones. I felt a little guilty for wishing that the blind was also equipped with a heater.

I heard them before I could see them, starting with a few loud hooting moans, clucks, and a strange resonant booming right outside our blind that soon became a full chorus. We were in the famous "booming grounds" of the greater prairie chicken, a traditional site where males gather each morning in spring to perform mating displays. Within ten minutes these wild cries in the darkness morphed into barely visible dark shapes scurrying across the grass.

Each male prairie chicken defends a small space, only about a metre across, where he performs a ritualized dance. These birds do look like a barred chicken at first glance, but when a male gets excited he raises the long stiff feathers that cover the sides of his neck and holds them erect above his head, sort of like bunny ears. This reveals bright yellow patches of skin on either side of his neck; the male stamps his feet rapidly, bows his head, and puffs up the yellow throat sacs like little balloons. Then he lets loose with a loud booming call that sounds like someone blowing on a conch shell. A male may repeat the dance again a few minutes later and dozens of time in a morning.

When it grew light I could see that there were twenty-five to thirty males gathered before us and about ten females in the audience. Males compete for centre stage, literally, because males who are able to control a display territory near the centre of the lek are preferred by females. Males begin challenges with a staring contest, where an intruder approaches, stands, and stares at his opponent. If one bird does not look away, or retreat, within thirty seconds or so the contest escalates into a kicking, pecking, and wrestling brawl. Despite the frantic activity, a female only rarely steps forward to invite a male to mount her. Females build nests and raise their families completely alone, and may spend many days at the lek before making a choice.

Jim had warned me we could not exit the blind until the birds had left the booming grounds. By 8 a.m. the sun was high enough that I imagined it was warming the air. The males were beginning to lose interest in their dances and only a few were still booming; the others were sitting quietly on the ground. Without warning, one bird took to the air and was immediately followed by the entire group, flying off across the prairie.

The show was over, just as suddenly as it had begun two and a half hours earlier. The next morning, the booming would start anew, drawing in female prairie chickens from miles away just as it has done for centuries.

~

Appearances are not everything, and in many birds the voice of the male is even more revealing to a female than his colours. Songbirds are well known for their impressive dawn chorus, a symphony of song that often begins before daybreak and dwindles to a whisper after sunrise. The best time for a female to judge a male's energy level and stamina is after a period of fasting, in other words, at daybreak. Singing is energetically expensive and females can judge male quality simply by listening to the quantity and quality of song.

In spring, silvereye males engage in a daily dawn chorus that consists of a complex song with a repertoire of up to sixty syllables per individual. Some males sing for less than fifteen minutes at dawn, while others go all out and sing for forty minutes. Males are relatively tame and tend to use the same trees as song perches day after day, which makes it easy to record their dawn songs.

Researchers at the University of Canterbury in New Zealand wanted to find out if a male's immediate access to food affected his dawn song the next day. Well-fed birds were expected to start singing earlier at dawn, sing longer, and perhaps even sing more complex songs. Each male's entire dawn chorus was recorded on three consecutive days—before, during, and after he received unlimited access to food for a day. Each fed bird was given a high-energy mixture of fat and sugar (think avian "cream puffs") from a feeder tied to a branch of a tree near his singing perch. Males sang the longest the morning after they were food supplemented, and their individual songs were longer, had more notes, and exhibited a broader frequency range than on normal days. The vocal muscles have to work harder to sing longer notes and produce a wider range of frequencies, so these are honest indicators of a male's energetic state. Well-fed males did not sing earlier in the morning, however, perhaps because a single day of extra food is not enough for them to shift their wake-up time.

Hooded warblers are small songbirds that nest in the undergrowth of towering forests in eastern North America. Males have a black hood that contrasts with a bright yellow cheek patch and throat, and they sing four to seven times per minute. My student Ioana Chiver caught males and used a hand-held spectrophotometer to measure the exact colours of the feathers, and then took a photo of each male so she could use the mug shot to measure the size of his black patches. Then she spent hours listening to each male to see how much time he spent singing, and how rapidly he followed one song with another.

To measure female mate choice, she radio-tracked females while they were fertile to see how often females visited neighbouring males and which males received the most visitors. Ioana found that female choice was influenced by a male's song rate but not his colours. A male who sang infrequently was more likely to have his mate sneak off the territory and visit neighbouring males who sang more often (Figure 4.1).

Most research on birdsong focuses on a particular group of birds that is specially equipped to produce intricate, complex songs that happen to appeal to our ears and sense of music. The oscine passerines, better known as songbirds, fill the spring air with a rich variety of songs that humans cannot easily imitate but can describe in simple terms. The red cardinal in the backyard sings *cheer, cheer, birdy-birdy-birdy* while the Carolina wren rapidly belts out *tea kettle, tea kettle, tea kettle, tea!* Though our ears allow us to appreciate the beauty, and to tell one species from another, birds produce and hear far more detail that is important in their own world of communication.

Birds have a complex sound-producing organ, deep in their throat where the airway splits into the two tubes that lead to the lungs, called a syrinx. Mammals have a completely different structure, called a larynx, which sits high in the throat. The syrinx of songbirds is controlled by an unusually large number of tiny muscles, allowing the bird fine control of the tension of the membranes that produce the pitch of the sound as well as the amount of air passing through the syrinx, which affects the volume. In some species, the songbird can independently control the two sides of the syrinx and create two different songs simultaneously.

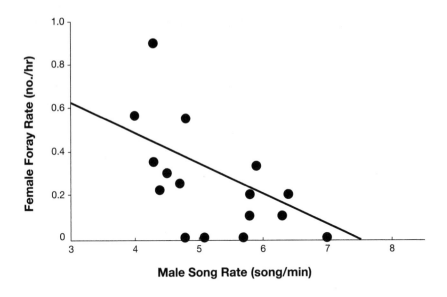

Figure 4.1. *Female hooded warblers leave their territories less often to seek an extra mate if their own mates sing frequently. The solid line reflects this trend. (After Chiver et al., 2008.)*

Though birds can see different colours than humans can, when it comes to sound we hear a similar range of frequencies. But birds are better at teasing apart sounds that come close together in time. What may be a single, sharp sound to us could, in reality, be three rapid-fire pulses that birds can discriminate. One way to appreciate the extra detail that birds hear is to play a sound at half-speed; what sounds like a beautiful song at normal speed is transformed into a drawn-out rendition with rising and lowering pitches, somewhat like the song of a whale.

Ironically, researchers study birdsong visually rather than by listening. Any sound can be represented as a computer-generated image, called a spectrogram, that shows the pace and exact frequencies of the sound energy captured by the micro-phone. A male blue-headed vireo, for instance, has a repertoire of five to eight different syllables that he strings together to form his song (Figure 4.2) and sings the same sequence repeatedly.

When near the nest, and approaching to take his turn at incubation, a male blue-headed vireo will repeatedly sing only one of his syllables to signal to his waiting mate. He picks the same syllable all spring, and even from year to year. Each male has a different repertoire of syllables but shares some in common with his neighbours. Males from distant parts of the breeding range may share few or no syllables and thus have distinct dialects, somewhat like a Texas versus a New England accent in human terms.

Blue-headed vireo song syllables are so distinctive that even a human ear can tell that the bird is changing its tune, but without recording and analyzing the song the subtle differences that birds can hear would remain concealed from us. Spectrograms allow a precise measurement of every detail of a bird's song that is not influenced by our own human perception. These images of sounds are used to tell if individual birds sing more complex songs or shift their frequency range and pace of delivery in response to social or environmental challenges. This is important, because with song the performer also broadcasts his health and genetic qualities.

The complexity of a male's song depends not only on his immediate energy stores, but also on his health and experiences while growing up. Songbirds learn their songs during the first months of life, and the details of how the brain is wired to remember and produce songs has been well studied in the lab. A songbird brain comprises special regions, located in the front of the brain, that are interconnected and control the learning, perception, and production of song. An individual's song complexity and size of repertoire in adulthood depends

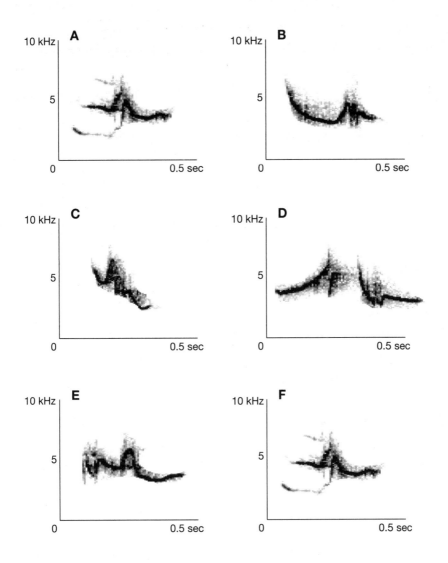

Figure 4.2. *Above are spectrograms of the song of a male blue-headed vireo. The song consists of several distinct syllables (diagrams A to E) sung in sequence, spaced about two seconds apart; then the sequence starts over again with the first syllable (diagram F). Note the different shapes of the syllables, each of which has its own unique changes in pitch (measured in kilohertz) over tenths of a second. (Images courtesy of Gene Morton.)*

on the size of these vocal control centres, which in turn depend on early development and, in some species, later experiences as an adult.

A male's song can even be affected by whether or not he had a big brother. The amount of steroids and nutrients a mother puts into her eggs changes over the course of laying the entire clutch. Testosterone levels, for instance, drop from the first to the last egg and can affect the development of the song centres in the brain. Some studies that have kept track of which bird came from which egg found that males from first-laid eggs are preferred by females. Why? An aviary study in Japan using a popular cage bird, the Bengalese finch, looked at laying order and song by taking early versus late eggs from different broods and letting a foster mother raise the chicks. The early-laid eggs turned into males who sang more complex songs, in that they switched often among a larger number of notes.

Exposure to parasites early in life also leaves its mark on a male's brain and subsequent song. In a lab study of canaries, researchers infected some juvenile canaries with avian malaria, a common blood parasite of birds. Once sexually mature, infected birds sang a smaller repertoire of song types and had vocal control centres that were half the size of those of uninfected birds. In canaries, very little song learning occurs in adulthood, so males cannot easily mask or overcome a handicap that arises in their youth. From a female's point of view, song complexity is an honest indicator of a male's prior exposure to disease.

Male European starlings defend nesting cavities and nest boxes, and in most populations about half the males are able to control multiple nest sites and have several social mates. These polygynous males are more aggressive and sing at a higher rate than monogamous ones. Helga Gwinner, a researcher at the Max Planck Institute for Ornithology in Germany, tested whether sons of polygynous males inherit these same traits of aggressive behaviour and higher song output. Once starling eggs hatched, Gwinner collected nestlings from both polygynous and monogamous males and brought them back to be reared in an aviary. Extra-pair mating is infrequent in starlings, so she assumed the male who owned the nest box was also the genetic father of the young.

A year later, the starlings were ready to breed for the first time and were tested for their ability to defend multiple nest sites. Gwinner placed one son of a polygynous father and one son of a monogamous father together in an aviary containing nest boxes, and allowed them to compete for a limited supply of nest sites. Sons of polygynous males defended more nest boxes and were dominant over sons whose fathers had had a single mate. Next, Gwinner placed two naive males in an aviary with a female; these hand-reared males had never before been with a female. Sons of polygynous males sang to females twice as much as sons of monogamous males, showing that both aggression and song output were linked, and could be inherited from the father.

A detective depends on evidence that has not been tampered with. The same is true for scientists who wish to test evolutionary hypotheses for bird behaviour. We glimpse into past processes by measuring the present-day traits and behaviours of male and females, and do experiments to see how displays, colouration, and song affect male mating success and female reproductive success. From that, we cautiously conclude that since females prefer showy males, and showy males are more disease-resistant or have more access to resources, then females who made such choices in the past must also have produced more high-quality offspring. In turn, males with high mating success also produced more offspring and those preferred traits would be favoured by natural selection.

The fundamental assumption of this evolutionary reconstruction is that present-day processes are similar to the processes that occurred many generations ago. Human impacts on the environment can change the behaviour we see in birds, alter selection pressures, and reshape the costs and benefits of a given behaviour. The link between present and past is forever broken, and history is lost.

An eager graduate student testing sexual selection theory might be puzzled by her results if she were studying female choice of male starlings in southwestern Britain, instead of Helga Gwinner's population in Germany. Our fictitious student would find that female starlings prefer males with a high song rate, but that these males have weak immune systems. How could females benefit from mating with males with poor ability to resist disease? Since there is no obvious answer, this study would likely get rejected time and again by the academic journal peer-review system that expects tidy results.

Unbeknownst to the student, her results do indeed make sense. Starlings in southwestern Britain are regular visitors to the rich feeding grounds at sewage treatment works, where earthworms thrive in the thick sludge left behind after the water evaporates. The worms and other invertebrates at sewage treatment plants in the United Kingdom and elsewhere are laden with environmental pollutants, including natural and synthetic estrogens. One of the more notable feminizing chemicals, biphenol A, made headlines after it was found in plastic baby-bottle liners and disposable water bottles. These chemicals mimic the real hormones inside an animal's body, and lead to abnormalities in the development of organs from the brain to the gonads.

Researchers at Cardiff University fed captive young starlings chemical-free earthworms, or earthworms tainted with a realistic dose of an estrogen-mimicking pollutant, or earthworms tainted with a cocktail of these pollutants. Males in the chemical cocktail group sang far more often, sang longer per bout, and had a repertoire size double that of males in the control group. This behavioural difference was a result of changes in brain development. Males that had been given female-like hormones had an enlarged song control centre in their brain. For all songbirds, the vocal centre is rich with estrogen receptors because early in development estrogen is critical for this brain region's development. The overdose from estrogen-mimicking chemicals, even for the year-old starlings in this study, led to larger song centres and thus more complex song.

Estrogens are immunosuppressants, however, so these highly vocal males also had weakened immune systems. In mate-choice trials, females naturally chose males with superior vocal performance because this strategy worked well in the evolutionary past and is now automatic. But in this contaminated population the choice backfires because male song is now a chemical indicator rather than a true test of male quality. Hormone-disrupting chemicals probably disrupt birdsong in a wide range of species, because these chemicals are used liberally around the world and are common contaminants in water supplies.

A male great tit flies to his female's nest box at dawn and sings vigorously until she comes out. This avian alarm clock, of sorts, is used by females to judge the male's stamina and health; female tits prefer to mate with males with a high song output and large repertoire. Researchers in Belgium studied great tits to find out if male song was affected by industrial pollution. They recorded and compared the dawn chorus of males living near a large metal smelter with those only 4 kilometres upwind. The smelter pumps out large amounts of pollutants and the lead and arsenic concentrations in great tit feathers were about twenty times higher in adults and nestlings there than in those at the upwind site. Adults in the two populations were in good body condition and nested equally successfully, but the lead- and cadmium-loaded males had muted voices. Males at the polluted site sang a dawn chorus that was 33 percent shorter in duration and they knew only two to four different songs, compared with as many as eight for birds at the relatively clean site.

It does not require an extreme example of modern-day pollution for bird brains to register and later broadcast the problem. Old enemies, chemicals that were banned years ago, are still affecting the sounds of nature. DDT was widely used throughout the world until the 1970s, when it was banned after the classic book *Silent Spring* made the link between this extremely persistent chemical and the disappearance of birds. DDT killed birds and disrupted reproduction by thinning eggshells to the point that they could not hatch; inside the body this pesticide disrupted a wide range of hormones. DDT reduced steroid levels in the egg and young bird, which in turn led to males having a smaller song control centre in the brain.

This was demonstrated for American robins nesting in apple orchards in the Okanagan Valley of British Columbia. Nestlings were taken from their nests and hand-raised, then kept in aviaries for two years. Males from orchards with high levels of DDT, and its breakdown products, showed a 15 percent loss in total brain volume and a 30 percent loss in the size of the song control centre. These results are especially alarming because male brains were not measured until the birds were two years old; the brain impairment was long-lasting and clearly the result of the brief exposure to DDT as an embryo and chick. Even worse, these effects were the result of DDT spraying some twenty-five years prior to the study! Perhaps we should be listening to the proverbial canaries in the coal mine, rather than waiting for them to fall off their perches.

~~◦

A different kind of pollution affects the singing behaviour of birds living in urban environments. In my suburban neighbourhood in Toronto, when I can finally sleep with the windows open in late spring, I hear the low roar of traffic noise from the distant highway. A dull rumble announces a plane going overhead, making its approach to the airport. Birds around the world have changed their song in urban areas so that their notes are not masked by the background human noise.

In this case it is not physiology and brain development that force males to sound different, but rather the built-in song plasticity needed for everyday communication allows males to adjust their tunes opportunistically in response to the noise pollution of city life. Songbirds have always had to compete for airtime with other birds and natural sounds, and males often modify their songs in different natural habitats. Background noise, whether natural or human-made, affects how far a bird's sound carries and the quality of the sound as discerned by the listener.

This topic is close to my heart because Gene, the birdsong expert in the family, was the first to show, several decades ago, that birdsong is fine-tuned for optimal transmission in different habitats. His PhD dissertation, "Ecological Sources of Selection on Avian Sounds," is a classic paper in animal communication. Gene found that tropical birds living in open habitats have high-frequency, high-energy sounds with short, quickly sung syllables so that the sound does not get as badly disrupted by heat and wind. Forest birds use low-frequency, long-wavelength sounds that are less likely to bounce off tree trunks and get distorted before reaching the listener.

In Sheffield, England, robins living in the noisier part of town sing during the night, rather than at dawn, to avoid the noise of the morning rush-hour. Nightingales living in Berlin sing louder than their counterparts living in the forest. Another way birds can make themselves heard in urban environments is to speed up the pace of their song and to shift to higher frequencies, the way birds normally do in open habitat. Traffic and city noises tend to be rumbling, low-frequency sounds, which would mask the low-frequency elements of a bird's song. From London to Amsterdam, great tits living in cities have eliminated the low-frequency syllables from their songs (Figure 4.3); they also sing shorter songs, and use unusual song types not found in forest populations.

The new phenomenon of city songs in birds most likely reflects plasticity by individuals and learning by juvenile birds, rather than genetic change in populations. Great tits, for instance, are known to adjust their songs according to their natural habitat and to copy songs that are used frequently by neighbours. They learn a large repertoire early in life, then can pick and choose from that repertoire according to local circumstances. Young birds who are born in cities do not hear low-frequency syllables during the critical song-learning phase, when songs are memorized for a lifetime, and so do not sing these songs when older.

In great tits, singing behaviour and flexibility in delivering notes evolved to fit their social interactions in their original

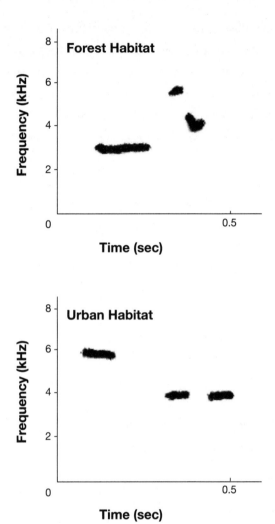

Figure 4.3. *This figure shows spectrograms of a great tit's typical three-note song recorded in a forest habitat versus one recorded in an urban habitat. Urban birds sing the first note at a much higher frequency so that it is not masked by the low-frequency background traffic noise. (After Slabbekoorn and Ripmeester, 2008.)*

forest habitat. Great tits can adjust to urban life because of this built-in plasticity. In this sense, they are just lucky that these pre-evolved features allow them to cope with man-made noise.

What is less well understood are the consequences of these city songs for male mating success and female reproduction. Males who adjust their songs to cope with city noises, so they can be better heard, may be breaking the connection between song characteristics and male quality. Large repertoires are often preferred by females, but if most urban males reduce the complexity of their song, then females may no longer be able to distinguish high-quality from low-quality individuals. Females of species who judge male song output during the dawn chorus may be unimpressed when males sing at night and are quiet at dawn.

Given how strongly song is influenced by a bird's internal and external environment, it follows that humans have caused sweeping changes in this feature of bird behaviour. This is not to say that cardinals will start sounding like they are asking for a cup of tea by singing like Carolina wrens, but our actions are changing the path of evolution. Females may choose males based on the wrong traits, in the case of toxins, which in turn erodes the species' adaptations for disease resistance that are normally linked to song performance. Urban noise may shift female choice to the point that, eventually, a new species is formed altogether because urban females will no longer pair with rural males. Will song sparrows, *Melospiza melodia,* living in cities someday be known as *Melospiza urbanii?*

5 'TIL DEATH DO US PART
Why Birds Divorce

T he wandering albatross has the largest wingspan of any
bird in the world, measuring about 3 metres from tip to tip.
Parents often travel 2,000 kilometres from their nesting colony
in search of food for their young and fly such remarkable distances by making special looping flights. As one wing tip brushes
the top of the ocean waves, the enormous bird banks into the
wind and quickly rises, then abruptly turns 180 degrees and
slowly descends until it can feel the ocean spray on its feathers
and holds its position. The long narrow wings are held out stiffly
on each side as the bird glides just above the water. An albatross
rarely has to flap its wings on its long trips, and uses the ocean

winds repeatedly to loop and glide. A parent may spend days commuting to rich feeding grounds where squid are plentiful before returning to its nest on a remote island in the southwestern Indian Ocean. Albatrosses are masters of the oceans and unconcerned with distances.

Their large ebony eyes search for distant flocks of petrels feeding on the surface; these small seabirds have an incredible sense of smell and are usually the first to find squid. Though food is plentiful in the ocean, it is hard to predict exactly when and where a shoal of fish or squid will rise to the surface. An albatross can easily bully its way into the middle of the flock and eat its fill. The partly digested squid and fish are stored in the upper stomach, and the oily, nutritious mixture is later regurgitated for the hungry chick. A bird returning from a feeding excursion is on the lookout for its nest, a raised bowl made of packed mud that contains a small white fluffy chick. The hungry chick rushes over and begins pecking at the parent's beak, flapping its stubby wings and begging loudly for its meal.

Though graceful in the air, albatrosses are much less so on land. The takeoff begins with a downhill sprint, wings held far out on each side, large webbed feet pounding the open ground and sand flying out behind every step. Like a jumbo jet lumbering down the runway, the albatross gains speed and finally lifts into the air. Albatrosses land awkwardly with wings outstretched, the chest sometimes hitting the ground to break the fall.

Pairs often greet each other after long feeding trips. A typical scene is a male returning to the nest and settling down on the ground to rest; his mate approaches and caresses his neck with her beak, gently pulling on his snowy white feathers. He reciprocates, and they begin to delicately preen each other's head, face, and neck. They may have been paired for almost twenty years, and would have shared the burden of many breeding attempts. The female must lay their one and only egg, of course, but they both take long shifts keeping the egg warm and travel enormous distances to find food for the chick, which will need their help for six months or more. The extreme demands of parenting require co-operation and commitment; albatrosses are truly monogamous and form lifelong pair bonds.

Choosing a mate is a long involved affair and divorce is almost unheard of. Albatrosses are usually eight years old before they are ready to breed for the first time, and females find a partner by visiting the display area where young males gather. Females waddle with a low rolling gait past several males and may stop in front of a male sitting on his empty display nest. The courtship begins with the female greeting the male with the usual ritual; she stretches her neck out low to the ground and points at him, quickly tapping the side of his bill with hers, then bows her head until it almost touches the ground and twists it slightly to the side so she can see him. The male then stands up and they begin the next stage of the display; both birds make volleys of loud, rubbery sounds as they rattle their beaks rapidly open and shut, slowly drawing their heads back and raising their beaks high.

The climax comes with the male's "sky call"; he stands with his white chest out and long wings fully outstretched and curved elegantly inward at the tips. Reaching high with his beak pointed skyward, he makes a series of rapid grunts and groans followed by a loud drawn-out braying scream. All the while, the female circles the calling male, forcing him to pirouette slowly with his wings open so he does not lose eye contact with her. Finally, they bow toward each other with a rapid head bob, and she walks away. This courtship display is repeated dozens of times, with slight variations, for some two years before a pair bond is finally complete and the pair breeds together.

Most birds pair monogamously but live with a given partner for only a few months or years, depending on the species. The pair bond, or "marriage" in human terms, is often a tenuous arrangement and divorce is a regular part of life in most birds. Annual divorce rates in birds range from 0 percent in the wandering albatross to 99 percent in the greater flamingo.

There are a number of theories as to why birds divorce their mates, each with differing predictions about who stands to gain and lose from divorce. The "incompatibility hypothesis" suggests that pairs who are genetically or behaviourally mismatched will divorce and both benefit from finding a new partner. The "better option hypothesis" is based on a conflict of interest between the pair; one pair member initiates divorce for selfish gain and leaves its former partner in the lurch. Finally, divorce is not always voluntary and may result from one pair member being evicted forcibly by a newcomer, often with the consequence that both of the original mates suffer.

～

Divorce has been well studied in the Eurasian oystercatcher, a large, chunky shorebird with an unmistakable appearance that is commonly seen along beaches and mudflats. The long, bright orange beak contrasts with the black head, bib, and back, and is highlighted by a thin orange eye-ring. The belly is white and the stout legs are washed-out pink. The hefty beak is used to hammer, stab, and probe at food including mussels, clams, crabs, small fish, and marine worms. Surprisingly, oysters are not a big part of their diet.

A colony on a Dutch island called Schiermonnikoog has been studied since 1985, a long-term research commitment necessary because oystercatchers live well over ten years. The annual divorce rate is about 8 percent, which means that each year a small percentage of pairs switch partners even though their former mates are still alive. During the breeding season a pair jointly defends a nesting territory as well as a separate feeding territory on the tidal mudflats. The best nesting territories are along the shoreline, adjacent to the mudflats, because chicks can follow their parents to the feeding grounds. The worst territories are inland, where chicks must wait at home while their parents commute to tidal flats and return with food. These differences in territory quality provide a strong incentive for inland oystercatchers to desert their mates in search of a home on the beach.

What are the consequences of divorce for the parties concerned? Individuals who desert a poor-quality territory gain a 20 percent increase in future reproductive success as a result of trading up—a modest but significant improvement. This strategy works best after bad winters, when a higher-than-normal adult death rate means there are widows and widowers in the market for a new partner.

In other years, the shortage of high-quality nesting sites means there is an incentive for birds on poor territories, or who have no territory at all, to fight their way onto a high-quality territory and force a divorce. These usurpations take many days of intrusions and fighting to accomplish. A bird cannot be evicted forcibly if its mate helps to defend the territory from the attacker, but in many cases the mate remains a passive bystander to the violence and does not interfere with the contest. Both the complicit mate, and the bird who forced its way in, will enjoy higher reproductive success, but the poor bird who was forced out produces half as many young in the future because it has lost its mate and prime real estate.

In many cases, oystercatcher divorce occurs voluntarily, at least from the departing bird's point of view. For the instigator of divorce, leaving a partner necessitates finding a new territory and forming a new pair bond—a risky strategy that pays off only if the bird can switch from a low- to a high-quality territory. The dumped mate, which most often is the male, experiences lower nesting success in the next few years compared to pairs that remain intact. Even a bird that retains a premium territory suffers from divorce because it has to attract a new partner. Newly formed pairs take longer to begin nesting, which causes a ten- to twenty-day delay in laying eggs, which in turns lowers reproductive success. This setback could be caused by a glitch in the fine-tuned coordination needed for a pair to jointly defend their territory, find enough food, and defend their nest from hungry gulls.

Resources, rather than the attractive characteristics or personality of a new partner, are also the underlying cause of divorce in common guillemots, cliff-nesting seabirds that often live thirty years. They form dense colonies numbering in the tens of thousands and feed off shore, flying under water using their wings for propulsion to catch fish and other prey. Guillemots have a stocky black beak, black head and back, white belly, and stubby tail.

At a colony on the Isle of May, Scotland, about 10 percent of breeders divorced each year during a twenty-three-year study. The guillemot lays a single egg on the bare ground, often on a narrow and precarious ledge. The egg is quite pointed at one end, so it tends to roll in a circle, but nevertheless many eggs careen off the cliff during bad weather or when the parents make a mistake switching places during incubation. Nest sites are not much bigger than the egg, and measure only 10 centimetres across, but are hotly contested, and pairs return year after year to re-nest in almost exactly the same place, less than a hand's width away. In large colonies, safe nest sites that are protected by walls and have a favourable slope are 100 percent occupied. At the Isle of May, the most productive ledges have been used almost non-stop, according to records from the 1930s.

Rather than being instigated by a pair member, divorce in guillemots was caused by aggressive birds that broke up the pair bond, studies show. When divorce occurred, the bird who was forced out of its rocky home suffered a lower breeding success because it had to move to a lower-quality ledge. The incoming bird produced more offspring than it had in the past, suggesting a strong advantage to aggressive divorce tactics.

In contrast to oystercatchers and guillemots, long-lived tropical songbirds are renowned for their permanent pair bonds and territories. Russ Greenberg and his wife, Judy Gradwohl, visited Barro Colorado Island, Panama, each winter for fourteen years to band checker-throated antwrens and map out their territories. Most years, the same bird was found living on the same territory, though individuals occasionally switched territories. When a bird eventually died and was replaced by a new mate, the territory boundaries remained unchanged because of the continuing presence of the neighbouring pairs. This is akin to taking a piece out of a jigsaw puzzle; if only one piece is removed, then only a same-shaped piece can replace it. This kind of long-term stability is not seen in migratory songbirds of North America and Europe because half the adults die each year; their situation is akin to taking out dozens of puzzle pieces at once.

My PhD student Sharon Gill, who studied mating systems in buff-breasted wrens in Panama, was also interested in divorce. The buff-breasted wren lives year-round in dense regenerating forest, often near rivers and streams. Pairs do just about everything together, including preening each other and singing duets. Buff-breasted wrens are famous for their antiphonal duet; they take turns singing individual notes, doing this so quickly that it sounds like one bird singing a long, complicated song. A pair sings together in territory defence, dueting at neighbouring pairs during border disputes.

During the four-year study, most individuals stayed with their original partner on the two dozen territories that Sharon followed closely. The briefest pair bond lasted only four days, after which the female was evicted by an interloper, and the lengthiest pair bond was four and a half years and was still intact at the end of the study. Mate switching happened during the dry season (January to March) when no breeding took place and occurred only in inexperienced pairs that had not yet bred together. After only five months of living together, and going through the trials and tribulations of nest building and tending eggs and young, a pair never divorced. These experienced pairs were broken up only through the death of one partner.

⁓

My first trip to Panama in 1995 to study antbirds was accompanied by a long list of dos and don'ts. I was used to working in a benign North American deciduous forest where my worst enemies were blackflies and poison ivy. In Panama, venomous creatures seemed to be lurking everywhere. "Don't touch anything!" seemed the most sage advice. The trunks of black palm trees are covered with long bacteria-ridden, needle-sharp spines—not something you want to lean against or grab for a handhold. Even after the palm falls over, the trunks litter the forest floor like a bungee trap if you happen to trip and fall.

There are other reasons not to grab trees and vines for support. I had never heard of the paraponera ant, or "bullet" ant, before going to Panama, but Gene had had the misfortune of running across one a few years earlier. Measuring about 2 centimetres in length, these ferocious ants are solitary and very territorial, attacking any moving thing that comes their way. Brush an arm or leg against a vine, and if you are unlucky, a bullet ant will strike and set you writhing in pain. Gene was laid up for two days in agony with an arm that swelled beyond belief.

On one of my first days in Panama, I was instructed to use my machete to clear a net lane for catching dusky antbirds. Not wanting to look like a sissy (and, I should add, I am pretty handy with a machete), I launched energetically into my job. Just as I heard someone say, "Watch out for that branch to your right," I gave it a good hard whack. Out came some unknown species of wasp, which proceeded to give me two painful jabs to the cheek— and I learned the hard way to look very carefully before hitting anything. But you can't always see the wasp nests. Once, Gene and I were walking into the forest to set up mist nets when a swarm of wasps suddenly flew out of the ground and stung him half a dozen times, and he was back in bed again with a giant, swollen arm.

During the dry season, we frequently stopped in our tracks to hear a loud buzzing sound overhead growing to a deafening roar . . . then fade away in the distance. This was a swarm of killer bees on the move, looking for a new home. Harmless, really, since they did not yet have offspring to defend. Still, walking up Pipeline Road one day, it was intimidating to find thousands of bees parked at eye level at the side of the road, a seething mass of venom the size of a basketball. I tiptoed away, even though I knew there was no real danger.

We were usually in Panama in the dry season, when it is hot, dry, and windy. I remember working our way down Pipeline Road, stopping the car to census birds along the way, and covering about 6 kilometres by late morning. Gene and I were standing in the forest playing back the songs of the dusky antbird, to see which banded birds lived on the territory. Although the wind was howling, we needed to do this last check before heading home. I didn't hear it, but Gene did. "Run!" he cried.

It took a moment for me to react, but Gene grabbed my hand and pulled hard. A gigantic branch crashed to the ground right where we had been standing, a victim of the dry-season winds. It was covered with moss, orchids, and large bromeliads, and had probably been aloft for many decades harbouring species of plants and insects not yet described by scientists.

Sometimes the attackers are entirely secretive. We had hosted a dozen Canadian students during a field course to teach them about tropical birds. The first week we had taken them to the Atlantic coast to see mangrove forest and the historical Fort San Lorenzo where the Spaniards stored their looted gold until it was shipped home. Two weeks after the class ended, Gene complained about his neck itching. Upon close inspection, I noticed a pinhole in a red swollen lump on his neck, and after some time saw a tiny white beast coming up for air. This was a botfly larva that had hatched out of an egg deposited on Gene's skin during our hike through the mangroves. Each time I tried to grab it with a pair of tweezers, it disappeared back into his neck.

It was time for home remedies. We had read that if you tape a piece of meat to the site, the larva will think beef is tastier than the victim and move from one meal to the next. After two days of walking around with a piece of filet mignon taped to his neck, he gave up on that. Next came the idea of Vaseline petroleum jelly; the larva needed to come up for air and so if we blocked off the air supply the thing would die. Gene walked around for several days with a Vaseline-filled beer cap taped to his neck, to no avail.

A few days later we were filling up the red Jeep at the only gas station in town, and Gene asked Obisbo, the tall black man who worked there, if he had any suggestions for getting rid of the larvae in his neck. As luck would have it, the driver of another car waiting for gas, a man who lived in a house at the top of the hill, told us we could drop by to borrow a chemical he used to treat botflies in his cattle. The chemical was a deep purple liquid, and seconds after the application the parasite was dead and easily removed. Gene was finally parasite-free.

And, as if that was not enough, there were the snakes. One of our most important study sites had been dubbed Snake Pit because highly venomous coral snakes were seen there regularly. Gene liked to tell stores about the bushmaster—an impressively long snake that can span a road and comes out at dusk to bask on the warm surface. The fer-de-lance is a pit viper, akin to the western diamondback rattlesnakes I had "met" many years earlier as a student in southern Arizona. One bite could be fatal.

I have not even mentioned tropical diseases, the price to be paid for entering one of the greatest evolutionary stages of life.

The diversity of plants, birds, and animals is at its highest in the tropics, and displays a bewildering array of adaptations to this unique and rich environment. Birds who follow ants, hawks who follow monkeys, ants who are fed by trees, wasps who live their entire lives inside fig fruits, gigantic damselflies who look like living fossils from the dinosaur era, giant caterpillars painted black and orange to warn off any animals stupid enough to attack, spiders that look like a bird's droppings—this is a biologist's paradise.

On a trip to Panama in 2007, when I was on sabbatical, Gene and I were walking down Pipeline Road with the kids, who were nine and ten years old at the time. It was only 9 a.m., but they were already complaining about the heat and humidity: "I'm tired" and "There's nothing to see." It is true that one has to be patient, and we hadn't seen anything new in half an hour. By mid-morning the birds were already hunkered down, the cicadas rising to a deafening roar, and the humans beginning to feel limp. Gene had turned back to get the car and I was trying my best to encourage the kids.

We turned a corner and I stopped cold in my tracks. "Look ahead, what do you see?"

Sarah continued to pout and refused to look. Douglas, sensing my excitement, spotted the birds. On the right hand side of the road, about a hundred metres away, we could see the rapid movements of dozens of birds, a hive of activity.

As we approached we saw an army ant swarm; hundreds of thousands of ants covered the forest floor like a living carpet. The antbirds perched within arm's reach of us, snatching up the insects that were fleeing from the ants. Almost every branch and leaf within a metre of the ground was covered with ants, all frantically scurrying along, but somehow the entire swarm, the size of a large living room, was organized and had well-defined boundaries. Douglas had read about army ants in children's books, but seeing the real thing was stunning, and he soon declared, "Thanks, Mom. I will never forget this moment for the rest of my life." Sarah, bless her, was unimpressed and just wanted to sit in the air-conditioned car when Gene finally drove up.

Our visits to Panama for seven consecutive years allowed us to band and identify almost all the dusky antbirds that lived along Pipeline Road. Dusky antbirds are heard more often than seen, and have a staccato song that increases in tempo and pitch (*da, da, da, da-da-da-da!*) and frequent call notes (*ta-dit, ta-dit*). Males and females are dead serious about defending their small pieces of the forest, and will charge at other dusky antbirds that dare to sing on their territory—and at our net, which has a playback speaker underneath.

About 90 percent of the banded birds were still alive when we came back a year later. The record for longevity was the "red-right" male who owned the territory near the Río Limbo for at least twelve years. He had been banded by another researcher before our study even started. Many pairs lived together for five years or more, but each year we found that several birds had switched territories and mates, so we knew that divorce occurred, if infrequently.

We wanted to see how mate switching actually happened but it was next to impossible to catch these birds in the act. So we created experimental vacancies and opportunities to divorce, by kidnapping an antbird and temporarily moving it indoors to an aviary. The captive bird seemed perfectly content with a liberal supply of mealworms, and even sang its courtship song from its small cage, as if looking for a new mate.

Back in the forest, the abducted bird's mate soon took a new partner. Most vacancies were filled within a day by a neighbouring antbird who suddenly divorced its mate in order to move in with the new "widow." This created a domino effect of territory and mate switching, where the vacancy created by divorce on the neighbouring territory was filled by yet another sudden divorce. These broken pair bonds were repaired when we eventually released the captive bird, who wasted no time reclaiming his original territory and mate, forcing everyone else to do the same.

The utopia of long-term monogamy in the tropical forests of Panama may result from lack of opportunity rather than heartfelt faithfulness. Dusky antbirds fit the pattern of apparently permanent pair bonds, but will divorce a mate without hesitation if a better opportunity arises next door. Faithfulness is enforced by long lifespan and tenure on adjacent territories.

~~~

Divorce has the most immediate consequences when it happens during the middle of a breeding attempt, leaving the partner single-handedly to complete care of the eggs and young, or worse, to abandon the breeding attempt and allow the offspring to die. If opportunities to re-mate abound, instead of finishing the job of parenting, a bird can boost its breeding success and save precious time by starting over with a new partner.

Snail kites are tropical hawks named for the food they eat. This hawk is among the most specialized in the world because it eats only snails and crabs. Its beak is needle sharp and very long and curved, a perfect tool for extracting the meat from a large apple snail the size of a child's fist. The wings are designed for slow flight, and are broad and rounded, so the hawk can fly low over marshlands searching for snails clinging to the vegetation. Male snail kites have a beautiful blue-grey body and a bright orange bill, eyes, and legs. Females are heavily streaked with brown.

Steven Beissinger, then a professor at Yale University, discovered that male and female snail kites are prone to divorcing their mates during the middle of the breeding season. Snail kites breed during the long rainy season, when snails are most abundant, and the male is a heavy player in parental care because he builds the nest, helps to incubate the eggs, and feeds the young. Beissinger first studied snail kites in the Florida Everglades, where they are an endangered species. His second study site was in the low, flat llanos of Venezuela that flood each wet season; here, a dozen pairs of kite may nest in the same tree.

Just over one-third of the females abandoned their mates and offspring even though there was a month of parental care still owing. Early abandonment allows the female to re-nest, something she would not have time to do if she stayed with her family and waited until her offspring were independent. And not only females deserted mates, as some 20 percent of males beat their female to the punch. This means the overall divorce rate was in the order of 60 percent or so, much higher than for seabirds or antbirds.

The kites most likely to practise divorce during parental care were those defending territories in areas with a high density of snails. Deserters were therefore not putting their offspring at risk, because there was plenty of food for the remaining bird to raise the half-grown offspring successfully. Beissinger tested the importance of food supply by manipulating the number of offspring in the nest. If birds deserted only when the survival of the nestlings was reasonably ensured, then divorce should be most common in single-chick families where it would be easy for one parent to provide all the food. Kites typically lay one to three eggs, so Beissinger added and removed young chicks from nests to create families of different sizes. The divorce rate skyrocketed to 90 percent in families with one chick but was totally absent in nests where there were three hungry nestlings.

In Italy, rock sparrow mates are also in a race to be the first to divorce. Rock sparrows are stocky sparrows that breed in rock crevices and nest boxes. About 25 percent of males desert their mates and brood, leaving the females to finish raising the young alone, while about one in ten females initiates divorce. Single-parent females increase their feeding effort substantially, matching the level of intact pairs, and so are able to raise young alone. Divorced males, on the other hand, do not (or can not?) compensate for the loss of help and fledge far fewer young.

Does the possibility of female desertion mean that male rock sparrows are alert to impending female fertility as the first nesting attempt nears completion. If a female is about to fly the coop to start a new nest elsewhere the male could leave first. To test this idea, researchers artificially weighed down some females, simulating the load of egg production, to see whether this would make males initiate a pre-emptive divorce. Loaded females, whether due to real eggs or equally heavy fishing weights (about 3 grams) attached to the base of the tail feathers, have a shallower ascent rate when taking off, a signal of fertility they cannot easily hide. A male mated to an apparently "fertile" female did not divorce her but instead enticed her to stay home. He spent more time near her, increased his courtship rate, and boosted his feeding trips to the nestlings.

The most extreme case of avian divorce occurs in the penduline tit, a small European songbird, where it occurs so rapidly that the association between partners can hardly be considered a pair bond. Male penduline tits attract a mate through song and begin the construction of a sophisticated, hanging, bag-like nest. The female assists the male with nest building (so far so good), but as soon as the female begins egg-laying, one of the birds abandons the relationship. The divorce rate is 100 percent and eggs and nestlings are always cared for by a single parent, most often the female.

Males have a head start in the divorce race because insemination must come before egg-laying. A female therefore does her best to conceal the onset of egg-laying from her mate by burying her eggs in the bottom of the nest. She also drives the male away when he tries to enter the nest to prevent him from looking inside to see if an egg has been laid. The rush to divorce is so intense that in 30 percent of families both the male and female abandon the nest simultaneously, dooming the eggs to sure death. Eggs are relatively cheap to make, and deserting them in order to re-nest, up to six times a year, can make a penduline tit who forgoes parental care vastly more productive.

I am often asked if checking nests and handling the nestlings causes the parents to abandon the nest. I used to say, "No, that's an old wives' tale." This myth, repeated by generations of mothers, discourages their children from bringing home cute little baby birds that are going to need an enormous amount of care and attention and likely won't survive anyhow. Young songbirds do look totally helpless and usually cannot even fly when they first leave the nest, yet they are being fully cared for by their parents, who are lurking in the bushes. I have checked hundreds of nests and handled even more tiny nestlings, and though the parents squawk and complain, they are not about to abandon several weeks of hard work. Once I go away, they cautiously return to the nest, find all is well, and resume feeding their young.

One exception caught me totally off guard. Gene and I started a study of blue-headed vireos in the mid-1990s at our forest in Pennsylvania and we banded the nestlings just as we have done thousands of other songbirds. When we went back the next day to check the first few banded blue-headed vireo nestlings, we were shocked to discover the little nestlings cold and dead in the nests. The males were nearby singing madly, but their mates were nowhere to be seen and never returned to the territory. A male does not brood the young at night to keep them warm, and so the young die if the mother deserts the family.

To continue the study and not harm the birds, we had to keep the parents ignorant of our visits to the nest. Before touching the nestlings, we caught both parents in a mist net and put them each in a paper bag. Then, with the parents effectively blind-folded, we picked up the nestlings, banded them, and returned them to the nest. A few minutes later we released the parents, and they went back to work never realizing that two large mammals had discovered their nest.

The mysterious question remained, however, as to why blue-headed vireo females abandon nests, but not hooded warblers, scarlet tanagers, American redstarts, ovenbirds, and all the other songbirds we have studied in our forest. Gene discovered the answer by radio-tracking female vireos near the time their young were due to fledge. Even without our interference, a female vireo starts making long-distance forays off-territory to wander the forest looking for a new mate. Then, an hour later, she is back on her own territory dutifully feeding her young.

At some point, a female disappears permanently, leaving her mate to care for the almost-fledged young, and for the next two or three weeks until they are truly independent. The male has no chance of attracting a new female while he is so busy, but meanwhile his former mate is fraternizing with a different male, already building a nest and getting ready to lay eggs that will be fertilized by her new beau. Our handling of the young had inad-vertently simulated a fledging event, since the nest was empty, prompting females who were already primed to divorce their partner to depart a few days ahead of schedule.

When it comes to divorce, birds are very practical and put their own personal interests first. Divorce occurs to increase access to resources and/or to produce more offspring than would be possible if a bird were to remain with one long-term mate. This is the stuff of natural selection—selfish behaviour trumps any thoughts of romance, or what might be best for the species as a whole.

~⌒

Albatrosses have been roaming the world's oceans for millions of years, and since long before humans became a common sight on the open seas. The rich ocean waters that have attracted seabirds for thousands of years now attract the longline fishing boats, putting albatross on a collision course with millions of razor-sharp, barbed hooks. Today, nineteen of twenty-two albatross species in the world are threatened with extinction. These spectacular birds have long inspired sailors, who admire their complete mastery over the harsh and often unforgiving oceans. The scientific name for the great albatross is *Diomedea,* named after the great Greek warrior of the Trojan War whose dead comrades were said to have been reincarnated in the form of long-winged seabirds who returned to fly beside his ship.

Tragically, the birds are completely unable to adapt their behaviour to overcome deaths caused by longline fishing. Each boat sets thousands of hooks on lines that are over 30 kilometres long and towed astern. The hooks, each the size of a child's hand, are baited with squid and weighted so the lines are dragged down to depths where swordfish and tuna can be caught. The smell of squid attracts a frenzy of seabirds who land in the churning water immediately behind the boat and try to grab the bait before it disappears below the surface. If its beak is hooked, the bird is pulled under water and drowns, and is retrieved when the fishermen haul up the line the next day.

When a breeding albatross is killed on a longline, there are dire consequences for its offspring and mate. After the egg is laid, the male takes the first turn at incubation and may sit there without food for many days, if not weeks, until the female returns from sea. If she does not come back, the male eventually abandons the egg to save his own life. If an adult drowns, the single young chick slowly starves to death because the remaining mate could not possibly supply enough food by itself. Widowed birds have to return to the mating arenas a year later and start over in the lengthy courtship rituals needed to attract a new mate.

BirdLife International, through its Albatross Task Force, teaches longline fishermen how to use simple measures that help save albatrosses, petrels, and other seabirds from being killed accidentally, the so-called bycatch of this fishing industry. Japanese bluefin tuna fishermen invented the "tori" line to prevent birds from stealing bait (*tori* means bird in Japanese). Working in the evening dusk so birds cannot see the bait, two lines of red streamers are set off the stern of the boat, on either side of the longline and its thousands of hooks. The streamers flap in the wind and discourage the winged thieves from coming close to the hooks. To make the bait hard to see under water, the pale squid is put in a large vat of blue dye before setting the hooks.

Fishermen are often the first to embrace these measures because they are the ones removing mangled dead birds from the lines and know that reducing seabird bycatch will mean more fish on the hooks. Many governments have passed laws that require monitoring of bycatch and enforcement of sustainable fisheries practices. In the Antarctic, for instance, the Patagonian toothfish longliners near South Georgia are closely regulated and careful bird protection measures have been hugely successful; the number of seabirds killed annually has dropped from more than five thousand in 1992 to fewer than thirty in 2005.

The intricate courtship displays and monogamous pair bonds of albatrosses are not a matter of choice, but rather have evolved over thousands of years as part of their ocean-going lifestyle. Lifelong pair bonds and a low reproductive rate are a necessity born of the difficulty of raising a chick in a harsh environment and the energetic demands of parenting. The wandering albatross is incapable of breeding each year, or suddenly laying two eggs to make up for its dwindling populations. It is equally incapable of learning to ignore tasty squid when it happens to be found near fishing vessels, so its future depends on human flexibility and our capacity for innovation and cultural shifts in behaviour.

6 YOUR TURN OR MINE?
How Birds Parent

It is understandable that there is so much conflict between the sexes over choosing and keeping mates, but there are also many tensions over raising a family. The interests of the male and female converge when it comes to being successful in hatching eggs and caring for young, so co-operation is essential. The conflict arises over who does the most work, a familiar tug-of-war in most families. The examples of divorce while in the middle of raising a family make the point, but there are also more subtle ways that a male and female pursue their own interests while caring for offspring together.

Even in monogamous species, males and females are often in

conflict because they do not stand to gain equally from taking care of the kids, for instance, when the male has been cuckolded. While this is of little concern to the female, who is virtually always the mother of the eggs in her nest, why would a male bother investing time and energy rearing someone else's offspring? The answer may be that a male cannot be sure whether or not his sperm, as opposed to someone else's, produced the offspring.

Hooded warbler males face this dilemma. About a third of females produce offspring sired by a neighbouring male, and a male hooded warbler makes over a thousand feeding trips to raise the offspring. If a male had some sense of whether he was on the winning or losing end of impressing his mate, he might adjust his parental care according to his likelihood of being the real father. All the newly hatched nestlings in a nest look alike: tiny, pink, squirming embryos that have lost their shells. A male certainly cannot tell an illegitimate baby from his own progeny.

I videotaped hooded warbler nests to find out just how devoted the male was to helping his mate bring caterpillars, flies, and other tidbits to the young. I was just as ignorant as the male about his actual paternity, but I had taken blood samples from the entire family for DNA testing. It turned out there was no relationship between a male's parental effort and his actual paternity. Some males fed four totally unrelated nestlings vigorously for over a week, and then took over as sole provider for another three weeks while the female re-nested. Other males, however, were the true fathers of their entire brood.

Female hooded warblers sneak off-territory to copulate with neighbours and are more likely to do so if their mates sing at a slow pace. One might think that a male could also judge his own vocal performance in relation to neighbours,' and therefore gauge his female's intentions. Males may not hold back on their parental care even if they suspect their mate of infidelity. Most nests that contained extra-pair young had mixed parentage; two or three nestlings belonged to a stranger, but at least one was the actual son or daughter of the male who was caring for them. Since a male cannot tell which one is his own baby, cutting back on feeding effort will hurt his own nestlings just as much as the others in the nest. A male's best option is to carry on as if cuckoldry never occurred, especially when the breeding season is short and a male cannot count on his female having time for another nesting attempt.

Infidelity could also influence male parental care from the cuckolder's point of view. Finding food for oneself and four rapidly growing nestlings takes time and effort that might better be devoted to sneaking copulations with neighbours. On the other hand, males could forgo liaisons off-territory and instead devote themselves fully to parental care.

My graduate student, Trevor Pitcher, radio-tracked male hooded warblers to see how often they snuck off-territory and he set up video cameras at all their nests to measure male feeding effort. The burden of parental care did not seem to slow male warblers down; they continued to spend some 10 percent of their time off their territory looking for females. A male can apparently do his fair share of parenting by matching his mate's feeding effort, about five trips per hour, and also have time to spare for searching out the few females who have lost their original nests and are re-nesting. Off-territory forays probably take less effort than feeding young, as males leave their territory only once or twice an hour and are gone for only a few minutes at a time. The payoff for these trips is potentially high because a male stands to double his reproductive output with relatively little effort.

In other species, males are caught in a bind between caring for the young and defending their turf against intruders. Blue-headed vireos are unusual among the northern songbirds I have studied in Pennsylvania because males help with almost every detail of nesting. The male must patrol and defend his territory, as usual, by singing his slow, crisp song from high in a hemlock tree. Watch for a bit longer, and he may switch to soft *yank* calls and join his mate in the understory, arriving with a bit of spider webbing, moss, or grapevine to add to their nest. Taking turns, each bird weaves its prize into the basket-shaped nest and then hops in to try it out, wiggling its body to custom-shape the cup.

Males incubate the eggs just as much as, or sometimes more than, the female. Not by coincidence, blue-headed vireos are genetically monogamous and so the male's efforts are not wasted on someone else's DNA. The pair carefully coordinates their arrival and departure from the precious eggs and rarely leaves them unattended. Nests with high levels of male incubation are more likely to hatch young because covering the eggs helps to camouflage the nest from hungry predators, and the parents are always nearby to harass blue jays, crows, and snakes that come too close for comfort.

This puts the male in a difficult position because he cannot easily defend his territory while sitting on the eggs. My student Ioana Chiver did her undergraduate thesis project on the male's dilemma by playing back vireo song on a male's territory—one time when he had just started his turn on the nest caring for the eggs and another time when the male had just come off the nest. Ioana randomly picked a spot 50 metres away from the male's nest, but within his territory, and did a three-minute playback of a song recorded from another male in the population, but not one within normal earshot. As doting and parental as a male vireo was, he responded instantly to hearing an intruder if he was not busy taking care of eggs. Within two minutes the male would be seen perched above the playback speaker, crown feathers raised in aggression, wings flicking in agitation, and singing up a storm of protest.

Forced to choose between attacking an intruder and tending the eggs, however, most males sat tight during the song playback. On average, it took incubating males almost twenty minutes to confront the intruder, as most waited for their mate to return to the nest before roaring over to the site where the singing originated. Almost half the males sang from the nest while incubating, providing at least a half-hearted challenge to the unknown challenger.

Under normal circumstances, when Ioana was not there doing playbacks, females spent half an hour off the nest during incubation breaks. During the experiments, though, females came back after only fifteen minutes and far ahead of schedule. Females seemed to be co-operating with their mates by allowing them off the nest early, but interestingly, females were still at least ten minutes late from the male's point of view. The playbacks began right after a nest exchange, so the female would have been hungry. Intruder, or not, she needed a snack before being stuck on the nest again for half an hour.

One of the more bizarre parenting arrangements occurs in a parrot species that lives in the rainforests of northern Australia. The greater vasa parrot has been dubbed the world's dullest parrot, since both sexes are solid black, but its mating and parenting system is anything but dull. Females are highly promiscuous and copulate with several males rather than forming a monogamous pair bond. Competition to fertilize eggs is intense among males, who have a penis-like structure that allows them to remain

Eastern Bluebird
A male bluebird is striking and would be even more so if we, like birds, could see ultraviolet light. The intensity of blue colour is a signal of quality; only males in good condition can grow the brightest feathers. Birds with vibrant plumage have a clear advantage in fighting for territories and attracting mates. (Photo by Marie Read)

Great Egret
The great egret's elegant courtship feathers were almost its downfall; this species was hunted to near-xtinction to decorate ladies' hats in the early 1900s. (Photo by Marie Read)

Wandering Albatross

This large seabird has a multi-year courtship (shown here is the sky dance display) and divorce is rare; pairs often breed together for twenty years. Almost all of the world's albatross species are endangered. (Photo by Paul Souders)

Long-tailed Manakin

Deep in a tropical rainforest, two male long-tailed manakins cooperate in an elaborate dance to impress visiting female (perched). Only the dominant male in the team gets to mate; his helper may have to wait years to inherit the display site. Note the coloured leg bands on the males, which allow researchers to tell them apart. (Photo by Marie Read)

Ovenbird

Well camouflaged in the forest, the diminutive ovenbird has a bold and distinctive song, "teacher, teacher, teacher," which it uses to attract mates and intimidate rivals. (Photo by Lang Elliott)

seate Spoonbill

s bizarre wading bird breeds in large colonies in the southern United States, Caribbean, and South rica alongside ibis, herons, and egrets. The distinctive pink feathers are a result of carotenoid pigments in l such as shrimp. (Photo by Juan Bahamon)

Blue-headed Vireo
The male blue-headed vireo helps its mate build the nest and performs ha... the incubation duties. Even when he hears an intruder singing nearby, the male stays put to protect the eggs. (Photo by Timothy Morton)

Tufted Puffin
This seabird nests in burrows on offshore islands in the North Pacific, and parents make long trips out to sea to find fish for their waiting chicks. Ocean warming has led to a drop in food supply, causing most colonies to crash in numbers. (Photo from JupiterImages)

coupled to the female for lengthy half-hour copulations. A male then continues to visit his mates, who are widely scattered, and feed the female while she is caring for eggs and young.

And this is not even the bizarre part. When the nestlings hatch, the female parrot quickly loses the black feathers on her head to reveal bright orange skin, leaving her looking somewhat like a vulture. She sits on prominent perches in the forest canopy squawking out an elaborate, raspy song. These songs can be heard a kilometre away by humans and are individually distinctive. Females who vocalize a great deal are rewarded with a high number of feeding visits by males. Males arrive from far and wide, perch by the female, and regurgitate a large mouthful of fruit for her to eat or deliver to her nestlings. Whether these males are her former mating partners is not yet known.

The vasa parrot arrangement for rearing young is very unusual and presumably evolved because the trees a female depends on for food are widely scattered and fruit at different times. The mating liaison allows her to enlist a small army of males who can search for food while she remains in the nest caring for the eggs. Feeding four to six young increases the pressure for food, hence the intense competition among females to attract males.

Why do males co-operate with female demands? Parrots are long-lived and known for their intelligence and long-term memory, so males who are good providers presumably gain copulations from the same female in future breeding attempts.

~⁓

Perhaps the greatest conflict in family life arises between the parents and offspring. Offspring face life or death depending on the effort parents are willing to make and risks they will take on their behalf. Young birds are vulnerable and usually expendable, and parents are used to having nests destroyed by predators and being forced to start all over again. The evolution of parental care can be thought of as a cold-hearted investment.

One spring I was birdwatching along the Texas coast and visited a huge colony of egrets, roseate spoonbills, and herons. Hundreds of large stick nests were built in the trees on a small island only 30 metres away, giving us a spectacular view. The air had the dank, sour smell of bird excrement and the wisps of off-shore breeze were most welcome. Half a dozen crocodiles lounged on the banks of the island, presumably making short work of any nestlings that toppled from the nests. At one nest, a magnificent egret stood preening its long white plumes. In the nest, a scruffy-looking nestling turned on its smaller nest mate and pecked its head, hitting it hard enough to knock it down. The parent continued preening, apparently unconcerned with the sibling violence. The entire time I was watching, about an hour, the larger nestling repeatedly and viciously pecked the other.

I was witnessing siblicide, where one offspring murders another, which is a common practice in herons. The parent did not intervene because it had created the situation in the first place by laying an "insurance" egg a few days after its first egg—to guard against infertility, which occurs when the egg is not fertilized, or accidental loss. A mother cannot easily feed and raise two offspring, yet routinely lays two eggs. When the first egg hatches the nestling gets a head start and grows fast. When it is old enough to do so, it dispatches its younger sibling using its large beak. The single child improves its own survival prospects, and the parents benefit too, because parental care is less demanding.

Parentally endorsed siblicide also occurs in many species of eagles, cranes, pelicans, and boobies, where broods of two chicks are always reduced to one. Even when food is plentiful, the older, stronger chick unleashes an unforgiving and lethal attack on its sibling. In other species, the fate of the second chick depends on the food supply the parents bring to the nest. In good years, the steady food supply dampens the aggression of the older chick and parents stand to double their reproductive success.

As all parents know, some children take more effort to raise than others. For birds, sons are often a burden because in many species males are physically larger than females, so they require more food. Common murres are long-lived, monogamous seabirds that nest on precarious ledges on towering seaside cliffs in North America and Europe. Males are only slightly larger than females, but because males wait on the cliffs for their mates to return to the breeding colony in spring, larger males have a competitive advantage over others. Heavier males can wait longer on the ledge, and thus are more likely to be present when the female finally shows up. Courtship is soon followed by copulation, and these pairs get a head start on nesting and are also more successful raising their young. Smaller males cannot wait at the cliff for a mate as long. These pairs feed their own young less often and are more likely to divorce.

Parent murres typically lose weight while caring for their single chick, an indication of the great effort required to fly out to sea, dive down 100 metres or more looking for a fish, and bring it back to the young bird marooned on the cliff. In one study, researchers from Memorial University in Newfoundland sat crammed in a wooden blind from dawn to dusk (sixteen hours) and watched deliveries to some thirty murre nests. They soon discovered that although mothers and fathers shared the workload equally, parents raising a son made 10 percent more feeding trips per day and lost weight rapidly compared with parents caring for a daughter

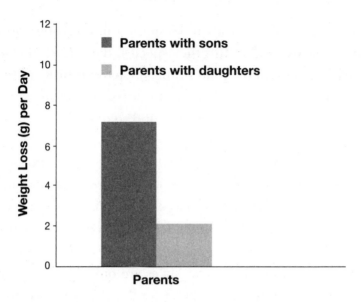

Figure 6.1. *This figure shows the daily weight loss of common murre parents of a single offspring. Parents of a son lost substantially more weight than parents of a daughter. (After Cameron-MacMillan et al., 2007.)*

(Figure 6.1). Raising extra-large offspring pays off only for sons, in this case, because they are likely to enjoy high mating success later in life.

The rhinoceros auklet is a sooty-grey seabird that nests in burrows, often on islands, along the coasts of the northern Pacific Ocean. The bird is not as ferocious as its name suggests, as the horn is just a small knob at the base of the bill that develops during the breeding season and then is shed. Long white feathers form whiskers and a crown at the back of the head. Large size is important in males because this allows them to defend a high-quality burrow. Since raising sons usually requires more work than raising daughters, parents should favour sons only in years when there is lots of food. This can be accomplished by providing more food and care when the single egg hatches out as a son rather than a daughter.

Rhinoceros auklets feed their single chick a diet of fish, and rely on rich, coastal upwellings in the ocean to fuel the marine food chain. The food supply changes dramatically from year to year and has a huge effect on nesting success. When overwinter sea surface temperatures are warm there is less food, and rhinoceros auklets breeding off the west coast of Canada are late breeding and only half the eggs hatch. Cool sea surface temperatures in winter increase upwellings and provide more food, so auklets breed earlier and almost all their eggs hatch.

In a year when environmental conditions are unusually good, most female auklets raise sons. Well-fed sons, who leave the nest in top condition, will be successful as breeders, but females can afford the extra investment only when food can easily be found. It is not known *how* females bias their family toward sons. Researchers could not determine the sex of the little auklets until forty-five days after hatching, when chicks were old enough to leave the burrow.

There is speculation that a bird can directly control the gender of her eggs at the time of fertilization, but it is more likely that her control is indirect and a consequence of body condition. In white-crowned sparrows, for instance, females with naturally high levels of stress hormones produced more daughters, as did those mothers who had been given experimental hormone implants. Elevated stress hormones in parents occur when food is scarce, and when transferred inadvertently to the egg can be especially harmful to larger male embryos. In good years, when food is plentiful, females may also lay eggs with more testosterone and fully equipped with nutrients, stacking the deck in favour of male chicks.

For the kakapo of New Zealand, sex ratio adjustment by females may stand in the way of protection from extinction. This is arguably the strangest parrot in the world, as it is flightless, largely nocturnal, and males are a third larger than females and can weigh in at over 2 kilograms. Males display at night from bowl-shaped hollows dug in the ground, and their loud booming calls attract females from far away. After copulation, the female leaves to build a ground nest, and she does all the parenting. A high-quality male can mate with many different females in a season, so sons can potentially produce many more grandchildren than daughters.

Kakapos are long-lived and females reproduce only every three to four years when the food supply is high. This low reproductive rate, combined with heavy nest predation by introduced mammals (for example, stoats, feral cats, and rats), has combined to make the kakapo one of the world's most endangered species. During the 1900s, kakapo numbers dropped steadily to the point that in the early 1980s there were eighty-two surviving birds, only a quarter of which were females. With such a strong skew in the sex ratio, and intermittent breeding, there were few females breeding in any given year. Those that did suffered nest predation.

In a desperate effort to save the species, the last kakapos were captured in 1982 and transferred to small predator-free islands. Twenty years later, the total population had declined further and many adults were decades old but had never bred on the island. Clearly conservation efforts had to focus on increasing productivity and boosting the number of female breeders.

Breeding usually occurs only in years when the trees in the native podocarp forest—with names like the rimu, kahikatea, and miro—all fruit simultaneously. The solution to reluctant females had seemed simple enough: provide high-quality food like nuts, apples, and sweet potatoes on female territories. The feeding program on the islands did improve female condition (15 percent heavier) and those females had more young, but the feeding somewhat backfired because well-fed mothers produced mostly sons. Mothers getting lots of food produced thirteen sons and only five daughters, whereas unfed mothers had mainly daughters (four sons, eleven daughters). The conservation program worked in that more females laid eggs, but the extra sons were of little use in the long run because, of course, they could not lay eggs.

To achieve an optimal sex ratio, the food was provided only after females had laid eggs, so that female condition would not affect the eggs. Two-thirds of the young from these females were daughters, and the females enjoyed high reproductive success owing to the food they were given during incubation and chick rearing.

⁓

Selfish behaviour predominates in most aspects of a bird's life, including choosing a mate, divorcing, and parenting. One of the biggest mysteries in the study of bird behaviour, therefore, comes when an individual shows unbridled altruism, showering help on neighbours at great cost to itself. In many bird species, parents are joined in their efforts by one, sometimes a dozen, extra adults who are temporary live-in nannies, or "helpers," and share the burden of parenting. In evolutionary terms, a bird that ceases breeding, even temporarily, will be underrepresented genetically in the next generation and therefore such a trait should gradually disappear over time.

So why is helping-at-the-nest so widespread in birds? Helping behaviour is especially common in the tropics and Australia and in certain groups of birds, like crows and jays. Among the superb fairy-wren of Australia, for instance, breeding pairs have up to four adult helpers living on their territory. These extra adults are always male, and help to defend the territory and bring food to the nestlings. In stripe-backed wrens living in Venezuela, helpers are always female. In South Africa, sociable weavers breed in colonies of up to two hundred individuals and build a large communal nest that contains multiple nesting chambers, each occupied by a different pair. Up to 80 percent of breeding pairs have male or female helpers, and some groups have up to five extra "parents" to raise the family.

Though it is usually clear how the parents benefit from an extended family (more food for the young and less work for themselves) it is less clear why a bird capable of breeding, in theory, would instead become a helper. Helpers must benefit in some way from spending so much time and energy as bird nannies. In other words, the altruism is not all it seems.

First, helpers may have a genetic stake in the offspring and thus are contributing genetically to future generations after all. If cuckoldry occurs, helpers could in fact be the real genetic parents of the young even though it might appear the helpers are not reproducing. Or, a helper may be genetically related to the parents and therefore actually be raising siblings, step-siblings, nieces, and so on. In this case, the genetic benefits of helping depend entirely on how closely related the helper is to the nestlings.

The first step in most studies of helping is to establish the true relationship between the helper and the recipient. With sociable weavers, genetic testing has shown that most helpers are former offspring of one or both of the parents receiving help. Helpers were not the actual parents of the young, because females rarely lay eggs in one another's nests and the dominant male is virtually always the genetic father of the young. A young weaver who cannot become a breeder can make the best of a bad situation by raising additional siblings and step-siblings. The gain in evolutionary fitness is diluted by the degree of relatedness; parents share 50 percent of their genes with their own offspring, while on average, half-siblings share only 25 percent of their genes.

Helpers may be especially important when environmental conditions are poor and parents cannot easily feed the young by themselves. Rita Covas, a PhD student at the University of Cape Town, found that among weavers, helpers had the greatest effect in years when rainfall and food supply were low. Then she added nestlings to some broods of sociable weavers to experimentally increase the burden of parental care. Nests with helpers should be better able to respond to the needs of enlarged broods than nests attended by pairs alone. Sure enough, nestlings in enlarged broods were fed more often, and survived better, when the parents had helpers.

In a hectic, bustling colony sociable weavers can somehow identify former parents and actively choose to help them instead of unrelated adults. How do helpers recognize close kin? For young birds who stay at home, and who up to that point have always been near their parents, individual recognition seems straightforward. In many co-operatively breeding birds, though, young move away from home and later return to help their parents. A study on long-tailed tits in Britain found that all adults attempt to breed independently in spring. Many nests are preyed upon, and although many failed breeders re-nest, some instead choose to help other adults that are already feeding young. Helpers preferentially care for close relatives, and predation is so high that for some individuals helping is their sole source of reproduction. Since helpers initially leave home to breed, there must be some way that, as adults, they recognize their parents a year or more later. In long-tailed tits, who do not sing much, recognition is via a *churr* call note that is learned while still in the nest.

Researchers at the University of Sheffield played back *churr* calls to pairs that were feeding nestlings, using calls from close relatives versus non-relatives. To find out what aspect of the signal was used for individual recognition they also played back calls whose maximum and minimum frequency had been digitally changed. Birds recognized the calls of relatives, but only if the calls had not been altered. Upon hearing a close relative near the nest, parents approached the source of the sound and gave their own contact calls, as a kind of greeting.

This showed that recognition occurred, but not how it developed in the first place. A cross-fostering experiment was done by switching marked nestlings between nests, then recording their *churr* calls a year later when they were adults. True siblings reared in different nests developed calls that were different from one another's, but similar to those of the adults who had fed them. Nestlings added to a foreign nest developed the same call structure as their foster siblings, showing that the subtle differences in the *churr* call are learned from the parents.

In birds like the sociable weaver, the parents clearly benefit from having helpers and the helpers gain reproductive success, indirectly, by bringing up relatives. But this raises the question of why helpers don't find their own mates and breed independently. A shortage of breeding territories could mean there are no vacant territories to occupy, so a helper's only choice is to not breed at all or to help relatives. In many cases, the helper eventually inherits the territory when the parents die, so may also receive a delayed benefit of helping.

Sociable weavers nest in colonies and new pairs do not have to leave the colony to breed. These birds live year-round in a communal thatched nest that is up to 6 metres across, and the colonies often have numerous vacant chambers. Weavers continuously add material to the nest mass, so a new pair could occupy an empty chamber or build its own. For the first two years, when young weavers are physiologically able to breed, most refrain from doing so and help parents instead. The proportion of nests with helpers varies greatly from year to year. In Rita Covas's study, only 30 percent of pairs had helpers in one year, compared with 80 percent the next year. Sociable weavers live in the semi-arid savannah of South Africa, where annual rainfall varies greatly and has a big impact on the bird's clutch size, and fledging success. In dry years, even established breeders may give up and lay no eggs.

Covas predicted that young birds would be less likely to help if there was a reliable and abundant food supply. During the breeding season, she provided a daily supply of canary seed to some colonies, and no extra food to other colonies. Birds enjoying the free handouts were heavier than those who had to fend for themselves, showing the food was in short supply under natural conditions. In the colonies without extra food, none of the yearlings built their own nests. In the fed colonies, however, 20 percent of yearlings attempted to breed and the average group size dropped because fewer birds were helpers.

Many studies have not found a clear-cut increase in reproductive success in groups with helpers, perhaps because the effects of helpers are seen clearly only in harsh years. Another possibility is that the parents shift the burden of care to the helpers and then hold back their efforts to improve their personal condition and survival prospects. Although no extra young result in the short term, this saved energy and effort on the parents' part increases their lifetime reproduction.

In Australia, superb fairy-wren nestlings receive 20 percent more food via helpers, but curiously this does not increase the mass or survival of chicks. The reason appears to be mothers that withhold investment in eggs if helpers are available. Assisted females laid smaller eggs containing fewer nutrients compared with females who raised chicks without help. The smaller eggs hatched into small chicks, but the helpers compensated for the female's under-investment so that the chicks grew quickly and caught up with their counterparts who had started life as large, well-nourished eggs. Mothers with puny eggs were more likely to survive to the next breeding season, which is important in long-lived species where lifetime reproductive success is heavily influenced by how many years one breeds.

The Seychelles warbler is named for the islands in the Indian Ocean where it lives; this species occurs nowhere else on the planet. Female Seychelles warblers go one step further in the subtle adjustments to parenting that have such a big influence on an individual's evolutionary fitness. This warbler has been made famous in the bird world by the series of surprising and detailed studies by Jan Komdeur, from the University of Gronigen in the Netherlands. Seychelles warblers, until recently, lived only on the small Cousin Island, where every individual can be marked and every territory is known. Territory quality is critical to parents' reproductive success, even though females lay a single egg, and also determines whether or not helpers are welcome. On poor-quality territories there is so much competition for scarce insect food that parents are actually better off not having helpers living with them.

The sex ratio of the nestlings also depends on territory quality. On low-quality territories about 80 percent of nestlings are male whereas on high-quality territories they are almost all female. In the late 1980s, there was no more room on Cousin Island for the growing population of Seychelles warblers, so some breeding pairs were moved to unoccupied nearby islands. Females that had been on a poor territory producing sons and who were moved to a high-quality territory immediately switched to making daughters. Hatching success was almost 100 percent, so the sex ratio of nestlings was not caused simply by embryo death in the egg; instead it must have reflected the initial sex ratio at the time the egg was fertilized, though no one really knows how females accomplished this feat.

The value of a son versus a daughter depends on a pair's territory quality because of the helping system. Sons always leave home and never help; pairs nesting on a poor-quality territory do not want helpers and so produce sons. Helpers are an advantage on high-quality territories, where there is enough food to go around, so producing daughters is a way to increase future reproductive success by ensuring a steady supply of help.

The parenting choices of birds are ultimately driven by the outcome days, weeks, or even years later. Evolutionarily the trick is to produce the most and healthiest offspring in one's lifetime through careful allocation of parental effort. Researchers measure parental effort by counting the number of food deliveries each parent makes, or by keeping track of the time spent incubating eggs, by watching the birds from a small, cramped blind or, if the action was recorded on video camera, from the comfort of a living room. The impact of parenting on health and longevity is much harder to judge.

Someone observing my parenting, for instance, might stake out my house and watch my many weekend comings and goings to soccer games, birthday parties, and the grocery store. Inside, surveillance cameras could measure my time devoted to maintenance of the home, food preparation, education, and instruction on appropriate communication skills. My efforts could then be described in terms of number of trips per day or percentage of waking hours spent parenting, though these objective numbers would not really capture my cumulative labour. An observer could nevertheless compare my statistics with those of my husband and draw conclusions about our respective work load and whether or not sons require more care than daughters.

Parenting is so costly that a bird's decisions are highly sensitive to environmental conditions. In poor years, when food is scarce, parents may produce daughters instead of sons, allow one nestling to kill another, forego breeding and redirect parental care to the offspring of relatives, or in the worst cases, abandon eggs and chicks altogether. Historically, this has served as an effective means of coping with the natural cycles that produce good and bad years. Today, widespread deterioration and loss of habitat, and global changes in weather patterns that affect when and where food can be found, are leaving their mark on birds' pliable and vulnerable behaviour and on their ability to produce the next generation.

7 EMPTY NEST
Finding a First Home

Eventually, young birds must leave their parents' territory and find their own breeding territory. In many cases, high-quality territories or territories of any kind are in short supply and young birds attain sexual maturity but cannot breed for lack of resources. One of the first studies to demonstrate the existence of a non-breeding population was conducted in the late 1940s, but the study was originally designed to answer the practical question of how important songbirds are in controlling spruce budworm. This native insect pest periodically breaks out and the larvae defoliate tens of square kilometres of spruce forest. The experiment seemed simple enough, though draconian: shoot all the songbirds in a forest plot and compare the budworm numbers with those in a comparable control plot where none of the birds have been removed. This experiment didn't work because the dozens of bay-breasted warblers, magnolia warblers, Cape May warblers, and blackburnian warblers that were shot were quickly replaced by young birds looking for a chance to breed.

My first study on birds, when I was an undergraduate student in the mid-1980s, looked at how young tree swallows find their first home. Tree swallows breed in tree cavities and have a delicate beak, so they must use old woodpecker cavities or nest boxes that humans provide. The short supply of cavities makes for fierce competition, and older females, who arrive first in spring, claim almost all the nest sites. First-year females are brown, instead of iridescent blue, and only a small number get to breed.

I was not a birdwatcher as a child, and neither were my parents, but I did love the outdoors, so as a college student I leapt at the opportunity to work outside all summer catching tree swallows and checking their nest boxes. In the spring I helped to set up grids of nest boxes in fields at the Queen's University Biological Station, near Kingston, Ontario, one of which was soon dubbed "Bridget's Grid." My job was to check the nest boxes twice a week to keep track of eggs and nestlings and to catch all the breeding adults. I learned how to take birds out of the mist nets that we placed in front of the nest boxes; there was so much fighting over the boxes that we caught dozens of swallows even though the nets were in the open and usually flapping in the cool May wind.

My favourite trick was feather tossing; male tree swallows will do just about anything for a luxurious downy feather. After installing nest traps that shut when a bird enters, I teased males by tossing up large white feathers that floated enticingly across the field. A male would swoop over and grab the feather in his beak, then, curiously, fly around the field with his gleaming prize to attract attention. Other males were soon in hot pursuit, often stealing the feather from the lead male in mid-air. The high-speed chase could go on for five minutes or more before one of the males finally delivered the feather safely home. The feathers were tucked neatly in the nest so that the broad ends curled up and over the tiny white eggs, keeping them warm and hiding them from predators when the mother was gone.

We caught the adults so we could give each one a small numbered leg band and to individually mark them. Unique combinations of coloured leg bands are one way to identify individuals throughout the season, and from year to year, without having to catch them over and over. Leg bands, though, are not easily visible on tree swallows because they have short stubby legs and when they fly the legs are tucked up, like an airplane's retracted landing gear. We added small daubs of acrylic paint to the swallows' wings and tail in different patterns, and with binoculars could then tell where a particular female (yellow, right wing tip) remained paired with her mate (blue, left tail) after her nest failed, and whether any bird had been evicted by a newcomer.

In many birds it is hard to estimate the number of non-breeders because they do not defend territories or sing, and therefore are difficult to count. In tree swallows, I counted them by giving the homeless birds a place to live. Halfway through the breeding season, when the territory owners were already incubating eggs or feeding young, I put up dozens of extra nest boxes in the hayfields that made up my study site. These boxes were quickly occupied by females who, until then, had had nowhere to breed; about three-quarters of the new owners were brown one-year-olds. Of all the yearlings I counted in the population, only 15 percent had been able to obtain a nest site naturally in head-to-head competition with older birds earlier in the season. Thus, in a typical tree swallow population most one-year-old females do not get to breed.

The traditional term for a non-breeder is "floater," implying that non-breeders literally drift from territory to territory and have little or no home base. Although survival of territory owners is usually quite high during spring and summer, once in a while one will be killed by a predator or accident. Floaters that wander widely increase their chances of coming across these rare, but valuable opportunities.

Young female tree swallows spend their time knocking on doors, so to speak, and wander widely in search of a vacant nest cavity. Sitting on a small hill at the edge of the tree swallow field, I could see the rows of nest boxes on their posts. Each one of the three dozen boxes was occupied by a breeding pair, and usually by an older, metallic blue female. Every ten minutes or so, I would spot a floater enter the field and begin to make her rounds. She caught my eye not because of her brown colour, but because she flew with her wings held down in an inverted V-shape, fluttering rapidly. The floater usually approached a nest box and circled it once or twice until scolded or chased by the owner, then she moved on quickly to another nest box. During the five to ten minutes that I could follow her movements, she visited half a dozen nest boxes to test if they were occupied. Floaters occasionally perched on a nest box, or even a nest entrance, if the owners were not present.

I set up an experiment to measure just how effective this floating behaviour was by catching older females that were building nests and keeping them safely in a holding bag for an hour. A young female usually arrived within ten minutes and started exploring the nest box. Imagine going to the grocery store and coming back to find someone has moved into your house! The intruder would land nervously on the entrance hole to peek inside, looking over her shoulder every few moments for any sign of attack. Then, if the coast was clear, she would begin going in and out of the box, and by the end of the hour would be chasing other hopefuls away. One female felt so much at home that she began copulating with the resident male. When the original owner was released she flew straight to her nest box and kicked out her replacement with little fanfare. First come, first served seemed to be the rule of the day, and the interlopers quickly retreated rather than risk serious injury.

Male tree swallows are in the same boat, competing with older birds for scarce nest sites. Young males are a beautiful iridescent blue like older males, so it is harder to count floaters versus breeders. We used a different approach altogether, and removed breeding males just after the young had hatched. A male without a territory or mate is desperate for a chance to breed, and we expected them to take over the nest box. This they did, within a day or two, but then newly arrived males were faced with the dilemma of caring for nestlings that they did not sire. Tree swallows have only one batch of young per summer, so although the male now had a nest box there was almost no chance he would ever be able to fertilize eggs laid by this female. He could simply wait another year to breed in this box, but the odds of living to the next summer were only about 50 percent.

Some newly arrived males took care of their problem by killing the small, helpless nestlings. We found nestlings dead in the box with peck marks on their heads. In other cases, the males picked up the tiny birds and dumped them out of the nest box while we were watching. The female had little choice but to pair with the interloper, because if she left, her odds of finding a new nest box and mate were slim. As violent as this is, the male had little choice but to force his new mate to re-nest.

For many years non-breeders have been thought of as floaters, but the tactics for home ownership are often more sophisticated than random wandering. In some species, non-breeders harass owners of surplus real estate to relinquish the extra resources; in others, they wait in line politely to inherit a territory, and in still others, they get downright nasty and steal homes.

My PhD project at Yale University focused on floater tactics in male purple martins. These large swallows are cavity nesters too, but remarkable because in eastern North America they live only in man-made apartment-style housing. I was interested in the female-like colour of one-year-old males that sets them apart from older males who have an immaculate and iridescent black-blue sheen from head to tail. The prevailing theory at the time was that a submissive colouration helped young males secure a nesting cavity within the colony.

My study site was the University of Oklahoma Biological Station, on the shores of Lake Texoma, which in those parts sep-arates Texas from Oklahoma. This was my first trip to the deep South and I encountered my first roadrunners, bull riders, and people who were astonished that I "believed" in evolution. Six purple martin houses stood on the front lawn of the field station, a perfect outdoor laboratory.

My first experiment looked at how one-year-old males acquire a nest site within the colony. Unlike the nest sites of tree swal-lows, which are always in short supply, the apartment-style hous-ing of martins allows many dozens of pairs to breed in a small area. The catch is that older birds, which arrive first, defend as many nest sites as they can even though they usually build only a single nest. To breed at all, young males first have to gain control of a surplus, but fiercely defended, nest cavity.

My job was to catch the young floater males and mark them with paint on the wings or tail so I could tell who was who. I used my old tree swallow trick of putting up housing to lure floaters, and raised small six-compartment martin houses on the lawn using guy wires. Within a day, a proud young male had claimed the house and was gleefully singing to attract a mate. Climbing a ladder, I installed nest traps in all the compartments to catch the bachelor. I banded and marked him, and then took down his house so he would have to return to the floating population and get a home the hard way.

Next, I needed to know how the female-like colour affected the male's ability to claim a nest site. So, adding further insult, I flipped a coin for every young male I'd caught and applied black Nyanzol D dye to half the group. This permanent dye, normally used for marking cattle, transformed the young males' white and brown female-like plumage to resemble the all-dark plumage of older males. Then I stood back and watched the colony with my telescope, looking for marked floaters to approach the nest boxes. When I saw one, I recorded how many territory owners he visited, how bold he was in exploring empty nesting compartments, and the reaction of the territory owner.

I learned that these purple martin "floaters" did not wander randomly among the nest sites, as did tree swallows. Each floater somehow settled on two to three older males to harass, and spent his day repeatedly visiting the same males. A typical encounter involved the young male landing on a porch and peering inside an empty nest cavity, and the older male charging at him to force him to fly off. A few minutes later, the youngster was back, cautiously checking out the empty compartment that the older male really did not need. After a few days or a week, the older male finally gave in and let the young male move in to one of his compartments. I was surprised to see that young males who had been given an adult-like plumage gained nesting compartments faster than female-like males. Older males yielded to the harassment more quickly if their challengers had adult plumage. So much for the supposed advantage of appearing submissive.

My next experiment was to see if the macho plumage of older males helped them to defend their nest sites from the many sub-adult males cruising through the colony. I visited a drugstore in southern Oklahoma and purchased five packages of hair lightener called "Born Blonde" (with my short, dark hair, I got some funny looks at the checkout). I caught adult males early in the season, before egg-laying had begun, and flipped a coin to decide whether the bird would be transformed into a pale sub-adult-looking male or remain as he was. My experiment showed that floaters did not make their house-hunting job easier by avoiding macho males and picking on other sub-adult or blonde adult males. I never figured out if there was any rhyme or reason to how young males chose their targets; perhaps they chose randomly, or saw signs of vulnerability in certain males that I could not see. Another loose end of my PhD study was explaining why young males looked like females. My colour experiments suggested there was no advantage to a floater, or to a territory owner, in looking dull.

One possibility is that the original advantage of a sub-adult plumage in a natural purple martin colony, with nest sites scattered between tree snags or in crevices of a cliff face, had been erased by the artificial housing. Purple martins are one of the urban winners in eastern North America, living almost exclusively in man-made housing in people's backyards. In southern Arizona, though, martins are still wild and live in saguaro cacti, using an abandoned woodpecker cavity for a nest site. In the 1800s and early 1900s, martins in the East nested naturally under boulders, in crevices or cliffs, and in tree cavities. Purple martins first encountered artificial nest sites when native Americans hollowed out dried gourds and hung them in their villages. European colonists copied this behaviour and built elaborate wooden houses for the martins. Intense competition for nesting sites, combined with the natural use of a wide range of cavity types, meant the martins were quick to adopt these new homes. Generation after generation, more and more young martins were raised in houses instead of trees. The simple rule of birds nesting in the same type of cavity that they were born in eventually would result in the entire population adopting houses in lieu of natural cavities.

~2~

My family usually spends the Christmas holidays at our farmhouse in Pennsylvania where we have a well-stocked bird feeder and plenty of customers. There is almost always a small flock of black-capped chickadees taking turns grabbing sunflower seeds, along with half a dozen other species. Sarah has been trying to teach the chickadees to eat from her hand. She lies down on the snow with her gloved hand outstretched and filled with sunflower seeds; I take the hanging feeder and put it down beside her hand. The chickadees are always first to the feeder in this potentially dangerous situation, scolding Sarah with a harsh *chick-a-dee, dee, dee* call. Before long, the other birds approach and Sarah is treated to close-ups of cardinals, tufted titmice, tree sparrows, juncos, and the occasional purple finch.

Chickadees have a bold look, too, with a jet black head and bib that contrasts with a bright white belly; they are hyperactive and rarely sit still for more than a few seconds. In winter they live in flocks, but in spring, breed in the nearby forest as monogamous pairs on well staked-out territories. Flocks have a complex social network and defend a large group territory from other chickadee flocks. When flocks encounter one another at territory boundaries, members give distinctive *tslink* calls and jostle and chase each other.

Within each flock is a clear-cut pecking order, with older birds in the top positions and socially dominant over younger newcomers. Low-ranked birds quickly leave the feeder if a more dominant flock member arrives, so rank determines access to winter food. Social rank in the winter flock is even more crucial for a bird's future breeding success; the top-ranked birds are paired even in winter and claim the best breeding habitat in the spring. Low-ranked birds often delay breeding for a year because there is not enough forest to go around.

Most young birds join a flock in their first winter as the lowest ranked bird and wait in line for a breeding position, gradually moving up in rank as older birds die. About 20 percent of chickadees take a different tack, floating between flocks and jumping the queue if a top-ranked chickadee disappears. When top-ranked birds are caught and kept in captivity for a few days, these highly valuable vacancies are filled not by the polite wait-in-line types but by a flock switcher. There is some level of social inertia in the domino-like moving up in the ranks, as there is little to be gained for a number-two bird in breaking its pair bond in order to move into the top position. This creates a glass ceiling of sorts and gives floaters an edge in the competition; it takes them only a day after a vacancy is created to go from rags to riches and guarantee themselves one of the best breeding territories and the most experienced mates. Chickadee floaters are discriminating, and completely ignore low-rank vacancies in the hierarchy because these are not likely to translate into breeding opportunities.

In some species, floaters are not actually homeless even though they do not yet control the resources necessary for breeding. Young birds may have a well-defined area where they live, but their space overlaps one or more breeding territories. So long as it remains out of the way, and does not attempt to mate or breed, the young bird can stay there safely. The advantage to the non-breeder of staking out its real estate from day one is that it will have first dibs on the vacancy over other non-breeders when the real owner dies.

Non-breeder territories are often much larger than what would be necessary for nesting, which allows the bird to wait in several lines simultaneously. In tropical rufous-collared sparrows, non-breeders form an underworld of territories that overlap the breeding territories of mated pairs. Young males and females do not compete for the same vacancy, and so each piece of real estate will have a breeder pair as well as a lurking non-reproductive male and female. Female rufous-collared sparrows are more likely to die in a given year than males, meaning that female non-breeders form shorter queues for territories by having smaller home ranges.

~⌒

When non-breeders live a wandering lifestyle, or live passively on the territories of breeders, they form a kind of secret society that can easily go overlooked. In co-operatively breeding species, though, it is obvious that many birds forgo breeding because they act as helpers and live openly with the territory owners. Co-operative breeding can be thought of as a two-step process that, first, requires delayed breeding and, second, involves the decision by the non-breeder to help the territory owners raise off-spring. The delay is caused by a shortage of good-quality breeding habitat, but the solutions are varied and can be sophisticated.

One of the more complex arrangements for finding a territory occurs in the Seychelles warbler, where some young birds leave home early to go it alone, while others stay and help, to boost their chances of moving in next door. Pairs defend territories year-round, and the main island, Cousin, is saturated with territories. About 120 pairs live on this island, and dispersal off island rarely occurs. Only 15 percent of adults die annually so vacancies for a breeding position are also rare. Young birds face a tough job finding a territory on the crowded island. Living at home permanently is usually not an option, because breeding vacancies are filled by a bird from outside the territory.

Young birds either become floaters and search widely for vacancies or stay with their parents and conduct their search from a home base by making brief excursions to other territories. Young birds are not always welcome at home, even though many help feed their parents' next batch of offspring. During the first year, young Seychelles warblers usually stay at home if both their parents are still living there. If one or both parents has died, and been replaced by a stranger, about two-thirds of the young leave home whether or not they have their own territory to move to. When a stranger moves onto the territory to take over the breeding role, it has little interest in the fate of the mate's offspring and may evict them.

Parents tolerate their grown-up offspring on the territory, despite food shortages, because this improves the odds that their kids will be able to acquire a territory elsewhere when they are older. Most young (70 percent) that are allowed to stay home are able to secure their own breeding territory within a year. Those that leave home prematurely are usually still non-breeders a year later, suggesting that floating is a risky strategy that often fails.

In the Siberian jay it is older siblings who do not tolerate competition, and aggressively evict subsequent younger birds from the parents' territory. This species lives in boreal forests throughout northern Europe and defends year-round territories. A group consists of the dominant breeding pair and up to four non-breeders that are former offspring or immigrants from other groups. A study in northern Sweden found that offspring often remain on their natal territory for up to three years, during which time they expel any new younger siblings that are produced. As a consequence of intense sibling rivalry, the ousted siblings travel up to 20 kilometres before resettling into a new group, where they too often wait three years before gaining a breeding position.

Non-breeder Siberian jays are clearly queuing for a breeding position because they do not help the breeding pair raise young. Parents not only tolerate the presence of their older offspring but actively defend them from predators. Offspring that stay at home during the first year have higher overwinter survival than young that are forced to leave the territory, and retained offspring also secure higher quality territories and thus enjoy high future reproductive success.

Offspring forced off-territory by their older siblings have no choice but to join another group, and by definition will be unrelated to the breeding pair and other group members. These dispersing juveniles are not welcome additions to the group because competition for breeding positions is already so intense. The dominant non-breeders aggressively chase away exploring dispersers, forcing the transients to move on to another group. The only place the desperate young can settle is on a territory where the breeders have a history of poor reproductive success, and thus there are no retained offspring defending a place in line.

The settlement patterns of Siberian jays have important consequences for the conservation of this species. Although dispersing young move long distances, they always join an existing group rather than breeding independently. Retained offspring and immigrants only move among neighbouring territories to occupy a breeding opening. Modern forestry usually results in extensive fragmentation of the forest and jays abandon intensively managed areas. Recolonization of isolated and unoccupied suitable forest patches will take a long time because jays only settle where other jays are already present, and as a result we can expect that populations will not recover quickly from logging.

The smell of smoke lingered in the air, and the cluster of pine trees in front of me looked like none I had ever seen before. The base of the huge trunks was charred black from the fire and the trees looked like they were bleeding. Rivers of white sticky sap had dripped down the trunks at a glacial pace, meandering around the thick flaky bark. Thanks to past and recent fires, I could see for hundreds of metres through this open park-like forest because there was no understory of shrubs and saplings.

I heard them before I saw them, a series of squeaky and excited *sklit sklit* calls. Two small shapes flew toward me, and then another that landed high on the tree trunk and began tapping on the wood. This was a red-cockaded woodpecker, whose name seems an exaggeration because the tiny red ornament ("cockade") behind the male's eye is usually not even visible. This spunky little bird has black and white barring on its back, a mostly white belly, and a jet black forehead and crown that contrasts sharply with the large white cheek patch.

Red-cockaded woodpeckers are the only North American woodpecker that drills nesting cavities in living trees, as opposed to dead tree snags. They are adapted to the southeastern U.S. pine forests where fire was (past tense) a normal and frequent event. Dead tree snags would have burned quickly, whereas trees like long-life pine are resistant to fire and provide a safe home for a woodpecker family. These woodpeckers excavate cavities only in old-growth trees, eighty years old or more, because the nesting chamber for the eggs must be in the heartwood where no resin will seep out and gum up the bird's feathers and harm its young. Some individual cavities have been active for decades and used successively by several generations of woodpeckers.

The specialization on old-growth trees, fire-influenced forests, and the difficulty of making new cavities has led to a mating system where sons, but not daughters, stay at home. Thus a typical family consists of mom and dad, plus several brothers or stepbrothers who help to incubate the eggs, brood and feed the nestlings, and clean the waste from the nest chamber. The rivers of sap on the outside of the tree are encouraged by the woodpeckers, who drill resin wells around the nesting cavity to stop tree-climbing black rat snakes from attacking the nest.

My guide that day was Nancy Jordon, who works at the Sam Houston State Forest in southeastern Texas. Her job, aside from endless paperwork, is to monitor the red-cockaded woodpeckers and to supervise the clearing of brush and managed burns. Red-cockaded woodpeckers are an endangered species in the United States, where less than 3 percent of the long-leaf pine forest remains intact. There are some six thousand breeding groups scattered in forest patches from east Texas to New Jersey.

Logging removed over 90 percent of the original pine forest, and the remaining forest has been managed to prevent fires. The result is forests that red-cockaded woodpeckers simply cannot use because they are choked with a thick mid-story of shrubs and saplings.

These woodpeckers are highly specialized in their habitat selection, and excavate only large, live trees. Their co-operative breeding system means that young females must leave home to find a new territory. Some 70 percent of females die during their search, a figure that is likely made higher by deforestation and the shortage of old-growth forest, which results in females travelling farther in their search.

Intensive rescue efforts require a detailed understanding of the woodpecker's breeding behaviour, not just the re-creation of the open grassy habitat under the forest through prescribed burns. Tree cavities are critical to the woodpecker group not just for nest sites, but also as safe places to roost at night. Groups with more high-quality cavities survive and breed better. Biologists have climbed ladders to drill holes into dozens of trees and placed artificial nest boxes in the tree trunk for the birds to use. Smaller holes are also drilled to get the resin flowing, and white paint is streaked onto the trunk to make it look like there is an attractive supply of sap.

Even with these efforts to create good habitat and a supply of cavities, however, young females have trouble finding these new sites, particularly if they are far from an existing woodpecker group. To overcome this obstacle, young birds are physically moved to jump-start new breeding groups in unoccupied habitat. A juvenile male and female are captured from a large group and moved together to the new habitat, which is equipped with artificial nesting cavities. And fingers are crossed that they will stay put.

This is a classic example of behavioural ecology informing effective management and conservation of a species very much at risk of extinction. Trying to save birds without understanding what makes them tick is a shot in the dark—providing suitable habitat is essential, but that alone is not always enough. Birds are highly social, and their social needs are at least as important as their physical needs.

8 FIGHT OR FLIGHT
Territory Defence and Aggression

The birds I study look fragile and delicate; warblers, tanagers, flycatchers, vireos, and swallows fit easily into the palm of your hand. Many are beautiful; a fire-engine red contrasting with black wings, a bold black and yellow face pattern, or even the more subtle but equally attractive steel blue-grey head with white eye-ring. They have small beaks for handling caterpillars and insects, feet designed for clinging to branches, and no outward sign of weaponry. Yet fighting and aggression is a daily event in these birds' lives as they compete for valuable resources like food, nest sites, and mates.

A long walk through the forest in the middle of summer would reveal little sign of actual fighting. A male hooded warbler sneaks onto another bird's territory to approach the female. Though he is quiet, the male who owns the territory spots him and flies in to perch 2 metres away, face to face. The staring contest begins, and both males fluff up their feathers and hold their wings slightly out and hanging down in a wing-droop posture. The intruder holds his ground, and the fight escalates to the next stage. The territory owner continues to droop his wings but points his beak skyward, showing off his black bib. Then, with no further warning, he flies at his opponent giving harsh *chippity-chup* growls and is almost quick enough to knock him off the branch. A frantic chase ensues, and the intruder is gone in a flash.

A male hooded warbler who sneaks around on other birds' territories gets caught one out of every five trips, on average. Most of these encounters end in a brief chase and a scolding, but occasionally an intruder is physically wrestled to the ground. Once, Gene was in the woods with his fancy reel-to-reel tape recorder and huge parabolic microphone as I had asked him to get some recordings of hooded warbler song that I could use to catch males. Gene has the most incredible luck, or more likely a sixth sense, for being in the right place at the right time with his recording gear. He had his parabola pointed at a male who was giving slow, threatening *chip* calls and spotted an intruder a few paces away. The owner gave a short chase and pinned the intruder to the ground on his back near Gene's feet, hammering away on his opponent's body and head with a small but very pointy beak. The intruder screamed for five to ten seconds, a frantic, squawking sound that I had never before heard, and finally wriggled free.

This rare recording of an actual attack sequence is now part of all our custom-made capture tapes that we use to lure male hooded warblers into our nets. Two minutes of crisp, ringing *wheeta wheeta weet-e-o* song on the tape usually brings the angry owner over, and he perches quietly high above the mist net checking out the situation. I set up a stuffed decoy of a male hooded warbler as a target, and when the tape explodes in *chip-pity chup* calls and screaming, the male loses his caution and flies directly at the decoy. If he comes from the wrong side of the net, he may actually land on the decoy and proceed to peck fiercely on the back of its head. I usually hide in the bushes on the decoy's side of the net, so I can flush the attacking male into the net if he gets too rough. These decoys were formerly live birds and are prized possessions in my tool kit that will last for many years if not ripped apart by a real hooded warbler.

Male–male chases are most obvious early in spring, when territory boundaries are being contested. Older males arrive first and reoccupy the same territory as last year, leaving newcomers to fill in the social gaps in the forest. This is more of a negotiation over where to put the fence, rather than an issue of catching an intruder red-handed near a mate, but some battles can go on for hours or days. Hooded warbler males chase each other in large circles near the boundary, perching often to sing and counter-sing. If one male switches song types, from *weeta-weeta-weet-e-o* to *weea, weea, weea,* then his neighbour will do the same. One flies at the other, and the circling begins again. The male who tires first will lose some ground, and after a few days will no longer contest the boundary.

The ability to win a contest depends on an individual's fighting ability, which includes weaponry and stamina, and its motivation to fight. The desire to win, or necessity, can trump smaller body size. During cuckoldry attempts, the defender potentially could lose his paternity of eggs laid by his one and only mate, whereas the intruder can come back another time and undoubtedly is visiting many different females. The defender has more to lose than the intruder has to gain, which is why intruders usually take flight at the first sign of trouble. Obtaining a territory is critical for both males and females, but the subtle details of exactly where the boundary is drawn will have a relatively minor impact on a bird's nesting success or survival. Boundary contests therefore are usually more subdued.

Differences in motivation to fight are clearly seen when territory owners are temporarily removed from their territories. If the territory owner has been gone only an hour or two, it can reclaim its territory from a newcomer just by showing up, and there is little or no contest. If a researcher allows the newcomer to become the owner, by letting it live there for several days, the contest for ownership is fierce after the original bird is released. The value of the resource is more similar between the contestants and neither one is eager to back down.

Gene and I did these kinds of removal experiments in Panama, to study territory switching and divorce in dusky antbirds. Even territory owners who had been sequestered in an aviary for two weeks were able to reclaim their territory quickly, something rarely seen in migratory songbirds. The difference, we think, is that tropical birds live on the same territories year-round for many years and vacant territories are harder to come by. A dusky antbird that has owned a territory for four or five years will fight harder to regain it than a hooded warbler that has spent only a few weeks on its territory.

In migratory songbirds, males are physiologically pumped up for fighting and develop high levels of testosterone for the short breeding season in spring and early summer (Figure 8.1). The intense rush to establish territorial boundaries against new

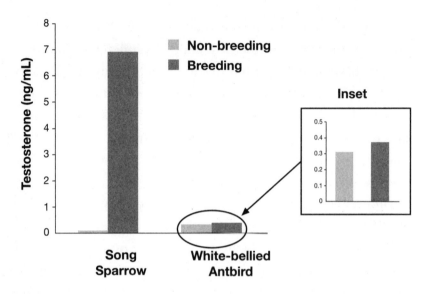

Figure 8.1. *Testosterone levels are very high in migrating songbirds (for example, song sparrows) at breeding time. Tropical birds (for example, white-bellied antbirds) consistently have very low testosterone levels, even when breeding. (After Fedy and Stutchbury, 2006.)*

neighbours and fend off sneaky intruders means the contests are frequent and the stakes are high. By mid-summer, though, the rate of challenges has dropped considerably and the male has a different priority—to feed the nestlings. Aggression and parenting do not go well together, and testosterone inhibits parental behaviour. Testosterone levels drop over the breeding season to practically zero, and rise again the next spring.

The physiological triggers for aggression are not quite so straightforward, because although tropical birds too are capable of extreme aggression, they have little or no testosterone circulating in their blood streams, even when breeding (Figure 8.1). My graduate student, Brad Fedy, studied white-bellied antbirds in Panama to understand how song and hormones are used in territory defence. These birds feed mostly on the ground and have long legs because they walk and hop so much, hence their scientific name *longipes*. Both sexes sing, though female song is most often heard as a duet overlapping the tail end of the male's song. The study showed that male testosterone levels were low year-round even though these birds responded quickly to simulated intruders (playbacks), sang back to intruders, and approached the playback speaker closely enough to get caught in mist nets.

Compared with temperate zone birds, tropical birds that live on the same territory all year contend with a low turnover of residents, so territorial boundaries are stable, and intrusions onto territories for sneaky copulations are rare. Singing rates, a reflection of male–male competition, are very low in tropical birds. White-bellied antbird males sing only a dozen times per hour, amounting to less than 5 percent of the time spent vocally defending the territory. A typical migratory songbird like the hooded warbler sings for 30 to 40 percent of its day, giving well over one hundred songs per hour. Chronically high testosterone levels are unnecessary for most tropical songbirds, given the infrequent fights, and would also be costly because testosterone suppresses the immune system.

Birds build social, rather than physical fences to keep unwanted visitors away from valuable resources. Underlying these invisible fences is always the threat of violence, and an intruder who crosses the line risks quick attack and injury. Most of the time, however, birds settle their fights with posturing, threat calls, and even rich, beautiful song. We have already seen that colour and song are important for attracting mates, but they are also used for sophisticated duels with neighbours to defend territorial boundaries.

~

Singing is a form of aggression that allows males to fight each other from a distance, sparring without actually exchanging blows. One of the most conspicuous vocal brawls happens at the crack of dawn, when males wake up and sing their hearts out. To the human ear, daybreak in summer is equated with birdsong, as forests and fields come alive with a beautiful chorus. Though pleasing to our ears, the songs we are listening to are actually arguments used to intimidate rivals and forestall attempts by neighbours to steal resources.

Threats are most effective if they can be backed up by real aggression. Singing costs energy, and at dawn males have had nothing to eat all night so are at a physiological low (imagine going on a 10-kilometre run on an empty stomach). Low overnight temperatures tend to decrease output of song the following morning, and food supplementation tends to increase song output during the daytime. Thus, a long or elaborate dawn chorus must be more costly and thus reveal male fighting ability.

The amount and timing of birdsong is not just a response to a bird's current energetic ability to sing, or an immediate response to an intrusion, but often falls into the category of preventive maintenance. Singing proclaims ownership of a territory and serves as a pre-emptive strike to dissuade neighbours and floaters from challenging the boundaries. Researchers work out the details of this communication with experiments that play recorded songs to birds to simulate a challenger and test theories of how birdsong is used in aggression.

In mild climates, male winter wrens defend territories and dawn-sing year-round. Males vary greatly in their dawn-song output because some males do not begin singing until shortly before sunrise. Males defend their territories vigorously even in winter, but a well-established territory owner can be expelled by a bold intruder. Researchers at the University of Basel, Switzerland, used playbacks to simulate territorial intrusions on wren territories in autumn. Since song is heard by rival males and prospective mates, it can be difficult to tease apart responses that are sexually motivated versus aggressive. Doing their study on males during winter, when no females were breeding, allowed them to focus on male–male communication. Each test bird was challenged by a male song one hour after sunrise, as though a new male had moved in. One day after the simulated intrusion, males increased their song output before sunrise, as if remembering the previous day's incident. Male wrens seemed to have learned from experience and appeared to anticipate possible future intrusions.

Birds go far beyond the simple expectation that hearing a rival will stimulate a territory owner to sing in defence. Songbirds often have individually distinctive songs and gauge their response according to prior social interactions with that bird, the likely threat it poses, and whether it is a stranger or neighbour. Not every song is created equal, and some song types are saved for the most aggressive interactions.

In France, researchers studied vocal battles in nightingale males by simulating two types of contests using interactive playbacks that allowed a researcher to sing back and forth with a wild bird. Nightingales are famous for their beautiful nocturnal songs, which males also use to attract mates. Songs were played at night, with researchers controlling whether playback copied and overlapped the songs of a target bird (mimicking an aggressive rival) or alternated songbirds with the target bird (as if the rival were not as big a threat). The next morning, males who had experienced an aggressive rival the night before were much more aggressive in territory defence than males who had been given a lesser challenger.

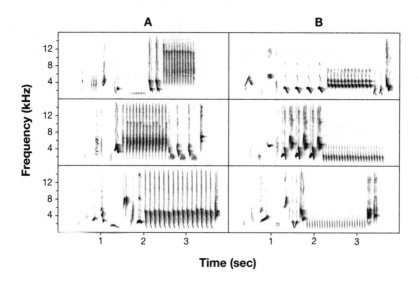

Figure 8.2. *Nightingale songs sometimes contain elaborate trills (rapidly repeated almost identical notes) of different repetition rates and frequency ranges. Broadband trills that span a large frequency range (A) are more difficult to produce in rapid succession than trills with a narrow frequency range (B). Males who sing broadband trills are more intimidating to rivals. (From Schmidt et al., 2008.)*

Not only do male nightingales remember strong challengers the next morning, but they assess the competitive ability of rivals based on a particular part of the song that is difficult to sing well. The trill used in nightingale songs comprises rapidly repeated almost identical short notes (Figure 8.2). The sound of the trills is a trade-off between how quickly a bird can repeat the individual units and the frequency range that each unit spans. Repeating identical notes in fast succession requires a precise coordination of vocal muscles and airflow and males vary greatly in their trill structure and pace. Rapid broadband trills likely indicate male quality and signal that the singer should be treated as a serious challenge.

Unmated male nightingales were exposed to two simultaneous playbacks early in the morning; one speaker sang rapid broadband trills in 50 percent of its songs and the other speaker sang songs without such trills. Territorial males focused their response on the speaker with the broadband trills, by quickly approaching that speaker and singing back to it at a high rate. In high-quality males, defined as those who subsequently attracted a female, the response was more aggressive to trills that had a broad bandwidth—meeting aggression with aggression. Males who remained unmated, and thus were likely low quality to begin with, had the opposite reaction. The broader the frequency of the trill, the more cautious the male was in his response; in other words, he was repelled by the aggressive-sounding intruder.

These playback experiments on nightingales show us how males fine-tune their song in a sophisticated way to sound more aggressive. It's not just a matter of singing a lot, but of interrupting the opponent during an argument and adding elements (broadband trills) that are meant to intimidate. Most songbird neighbourhoods are social networks; a territory owner recognizes his rivals individually, and continually updates his assessment of each rival's immediate and future threat. This allows the territory owner to give a response that is in its own best interest: be aggressive if you are not bluffing, and retreat if you are.

～✺

Song is a dynamic form of communication that can be modified from moment to moment, and from day to day. Though it is constrained by an individual's physical condition and upbringing, it functions well at close range and over long distances when contestants cannot even see each other. Birds also communicate visually via their plumage, which is used primarily for short-range contests. Feathers are regrown only once or twice a year, so colour is a more static form of aggression and reveals the bird's condition when the feathers grew.

Feather postures, on the other hand, are an instantaneous tell-all. A nervous bird will hold its feathers sleeked back, close to the body to look small and submissive, whereas an aggressive bird will raise its crest, fluff its feathers, and hold its wings out to make itself look intimidating. Still, a bird cannot change its feather colour at will. Colouration, just like song, is effective in communication only if it intimidates opponents and allows contests to be resolved without resorting to dangerous fights.

Underlying aggressive behaviour, at least in migratory birds, is testosterone; this has been shown repeatedly by implanting males with testosterone and observing an increase in singing and territorial patrolling, and more success in winning contests. The question, though, is whether a male's colours signal his high levels of aggression. House sparrows, for instance, that are given an extra dose of testosterone grow larger black bibs and become socially dominant over normal males.

Dark-eyed juncos have been studied extensively by Ellen Ketterson's lab at Indiana University to dissect the details of how testosterone influences a male's territorial and mating behaviour. Male juncos are dark grey on the back but have brilliant white outer tail feathers that the birds spread in both courtship and aggressive displays. Males whose tails were painted whiter became socially dominant and were more attractive to females. Males captured immediately after a simulated intrusion (a playback tape) had higher testosterone levels if their tails were naturally very white, showing that this was an honest signal of a male's readiness to defend his territory.

In many species, an individual's plumage is highly variable, with multicoloured patches on its body that could serve different functions in communication. A hooded warbler has a bright yellow cheek patch and breast, a black crown and bib, and white outer tail feathers. A northern cardinal male is solid red, but males have a prominent feathered crest that can be raised and lowered. Which of these "badges," if any, are used in male–male contests?

The lark bunting breeds in the grasslands of the midwestern U.S. Females are camouflaged in dull browns, but the male has a jet black body with large white wing patches. Males arrive on the breeding grounds a few weeks before females, so for a short time communication is clearly among males and does not involve female assessment of male quality. Social interactions between males range from benign associations while foraging off-territory, to establishment of territory boundaries, to fierce defence of a female against marauding groups of males (called mobs). With such a rich variety of contexts in which males compete head to head, and given the sharp contrast in appearance between males and females, it is likely that the size and detail in male colouration has some function in aggression.

At the Pawnee National Grassland, Colorado, unmated lark bunting males were presented with a simulated intruder—a decoy perched in the middle of the open territory that dares to sing. Decoys with large white wing patches received less aggression from territory owners, but decoys with more black were approached more quickly and received more aggression than lighter models. This opposite response to the two colour patches was puzzling at first, but reflected the different "jobs" of these plumages.

During natural observations of males, males with large white wing patches received fewer intrusions and challenges. Since the white wing patch contrasted sharply with the black body, and males perched conspicuously on shrubs in their territory, the size of the white patch could be seen at greater distances and was used to avoid conflict. Winners of contests had a larger proportion of black feathers on both rump and body than the defeated birds. The extent of black colouration is effective at short range, perhaps because the amount and intensity of black can be assessed only when individuals are physically close together. The multiple badges, black body, and white wings are not redundant signals but instead work in concert at different ranges and in different social contexts (before and after intrusion).

The Toolangi State Forest in Victoria, Australia, is home to the golden whistler. Males do not attain full adult plumage until their third year, and even then often retain traces of juvenile plumage. The colour of fully mature males includes a white throat patch, yellow breast, and nape band, and a black chin-stripe. During natural close-range encounters between rivals, male golden whistlers perch near each other and display their throat patches. Researchers at the University of Melbourne found that males with large white throat patches defended large territories.

To find out how the white colour affects male social interactions, they captured males and reduced the size of the white throat patch using a permanent black marker. Other males were captured, but sham-manipulated so they received the same amount of handling but had water applied to their feathers, and looked no different afterwards. Males were housed in an aviary, and taken out in small cages, one at a time, to simulate intrusions on real territories. Defending males directed higher levels of aggression toward caged "intruders" with large throat patches, responding for longer and spending more time near them. By manipulating the throat patch directly, researchers could isolate the effects of this signal versus the more colourful features of the bird, which, as it turns out, did not affect the outcome of contests.

One of my first experiences as a bird detective, when I was an undergraduate student at Queen's University in Ontario, was the discovery of two female tree swallows fighting inside a nest box. One female was clearly on the losing end, because she had been scalped; the feathers on the top of her head had been stripped off completely. I was shocked at the extent of injury that these beautiful little birds could inflict on one another, and that it was two females locked in battle.

Most studies of aggression, song, and badges have focused on male–male fights, in part because these are so conspicuous in most birds. In many species only the male sings, and he is far more colourful than the female, so naturally most research focuses on male contests. Male fighting is more observable because male reproductive success is determined primarily by the number of mating partners he can obtain, and a near unlimited supply of sperm means the sky's the limit. Female reproductive success is determined most by the number of eggs she lays and the quality of her offspring, so resources for breeding rather than the quantity of partners is most important. Females may compete for high-quality territories or males willing to provide a high level of parental care. In cavity-nesting birds, like tree swallows, females cannot breed unless they can control a good nest site, so female–female agression is extreme.

In species where males routinely have multiple mates living on the territory, the resident female benefits from preventing settlement of other females with whom resources would have to be shared. The bluethroat is a small migratory songbird that nests on the ground in willow meadows in southern Norway. These birds are typical in that males arrive in spring before females, and use complex songs and flight displays to defend territories and attract mates. Females select mates by entering the territory, viewing the male's vigorous head-up posture and raised tail, and listening to his songs at close range. Males are sometimes polygynous and attract two females that nest on the territory. Females are generally secretive during the breeding season, but also sing to try to keep other females from sharing the territory.

To study female–female aggression, researchers placed a caged female bluethroat, accompanied by female song playback, on territories in which females had not yet begun laying eggs. When challenged, female territory owners came out of hiding, approached the caged intruder, and sang at it. Sometimes a male showed up, and if his mate was not around, displayed to the female decoy as an invitation to move in. Female aggression was muted compared with male responses to male decoys; a male presented with a caged rival approached faster and more closely, and flew over the cage many times.

Not only are studies of female aggression few and far between, but also little is known about what controls aggression inside the female's body. Testosterone is sometimes associated with female aggression in birds, but operates at a much lower level in the blood stream compared with that of males. Even after a simulated intrusion with a live bird, for instance, male bluethroats had testosterone levels over ten times higher than females. Estradiol is the most potent, naturally occurring type of estrogen that is thought of as a female hormone (though it also occurs in males), because of its important role in the growth of sex organs and the menstrual cycle of humans. Estradiol may also play a role in female aggression; female bluethroats that responded strongly to an intruder tended to have higher levels of estradiol.

Female song is rare or absent in many species, but females nevertheless are vocal in their territory defence by using simple call notes to keep intruders at bay. I have spent many mornings at our farm in Pennsylvania, walking through the nearby forest listening for the *chiff* calls of female Acadian flycatchers. During nest building and incubation, females call almost incessantly while feeding, gathering nesting material, and taking a break from incubation duties. Fertile females spend about half their time chiffing, and on a good morning I can find half a dozen nests during a casual stroll up the dark, hemlock-filled stream valleys. Though the sound is not nearly as obvious as the male's explosive *peet-za* song, female chiffing is clearly audible from 100 metres away. When silent, however, the dull-coloured female is near invisible in the boughs of the hemlocks overhead.

My graduate student Stephanie Hung compared female Acadian flycatcher behaviour before, during, and after playbacks of female chiffing. Though females quickly approached the playback speaker to find the intruder, we were surprised that they did not confront the intruder by elevating their own rate of chiffing. Instead, females switched calls and gave a sound we rarely had heard during our many hours of radio-tracking and observing females. When challenged directly, females gave high-intensity *wheuu* calls, about every five seconds. Female Acadian flycatchers use chiffing as a general deterrent to announce that the territory is occupied, and reserve their most aggressive call for dealing with intruders. Females who are not aggressive risk being kicked off their territory, or may even have to share their territory with a second female.

The skew in armaments, ornaments, and aggression between males and females occurs because of different sex roles in establishing territories, mating, and parental care. When sex roles are similar we expect to find male and female are equally well equipped to fight, both physically and behaviourally. Many of the tropical passerines that I have studied in Panama fall into this category because both sexes sing, both participate in nest building and incubation, cuckoldry is infrequent, and the pair lives together year-round.

White-bellied antbird pairs in Panama defend their territory through song, but sing only once every five minutes. If you sat on their territory for one-hour shifts, like my student Brad Fedy did, you often would hear no songs at all. Rarely, one hour out of fifty times, you would happen to be on the scene during a dispute, and the singing would be almost non-stop (two hundred songs per hour). Males take the lead in singing, in terms of both giving more solo songs than females and leading the pair in a duet. Playbacks showed that males responded faster and sang more than females in response to intruders, and pair members responded equally strongly to playback of duets, male song, and female song. It was unusual that females responded aggressively to male song, and vice versa, and this suggests some level of co-operation in territory defence. Temporary removals of a pair member showed that widowed birds did not immediately advertise for a new mate and vacancies were not filled rapidly, perhaps because pairs benefit from remaining together rather than being the victim of a forced divorce.

Sex roles are identical for species that defend individual territories outside the breeding season. Many migratory songbirds, particularly those that eat insects, defend a territory on their tropical wintering grounds to guarantee a steady supply of food for the next six months. There is no co-operation at all between the sexes as each bird must find its own territory. Females compete not only with other females, but also directly with males. Since there is no mating or reproduction, singing is largely absent and territorial boundaries are established and maintained via call notes that both sexes share. On this even playing field, males do have the upper hand by virtue of their larger size. Females tend to settle into secondary, poorer quality habitat while males claim territories in the wetter, food-rich forested areas.

My first field study on the migratory hooded warbler was in a large forest near the Yucatán coast, south of Cancún, Mexico. This dry tropical forest was a far cry from the lush, deciduous forest in Pennsylvania where from one spot I can hear half a dozen males singing. In the Yucatán, song is reserved for year-round residents and migrants like the hooded warbler very much blend into the background with innocuous, metallic *chip* calls. Though simple in structure, and repetitive, the call serves its purpose of proclaiming ownership.

I lured hooded warblers into my mist net by playing back chip calls, and mapped out whether a male or female owned each territory. Most females had shared boundaries with males and I wanted to know how difficult it was for females to defend their territories against males, and whether the black hood of males was used to intimidate females. One-year-old female hooded warblers have no black signal at all, but older females range from having a black outline on the crown and breast to looking almost like males. Perhaps this black badge is used by females to threaten other females, and hold off males, during competition for winter territories.

My experimental removals of territory owners early in the fall in Mexico resulted in rapid replacement by nonterritorial "floaters." Replacement birds were not necessarily the same gender as the original territory owner but virtually all replacements were young birds facing their first winter in Mexico. Thus, females were able to acquire and defend territories in the presence of male floaters.

Next, I staged intrusions by putting hooded warbler decoys in the middle of a male's territory—presenting him with a yellow female, a black-hooded female, or a male. Males approached the model giving *chip* calls, and many perched close to the model giving the "wing droop" and/or the "throat up" display. A few lost their cool and hit the model, and males attacked both female and male decoys. My results suggested that male-like colouration in females does not give them a competitive edge with males. It is still a mystery to me why some females are so male-like, but I've never been able to talk any of my new graduate students into following up on this tricky question.

Intense competition for resources among females ramps up levels of aggression and the need for aggressive signalling to settle contests. A minority of bird species feature so-called sex-role reversal, where females become the more aggressive sex due to competition for mates or nesting sites and males take on the domestic duties by building nests and caring for eggs and young. In such species, females are usually larger, brighter, and more aggressive than males.

In northern Australia, the eclectus parrot features dramatically different sex roles in competition. View a pair perched side by side, and you'd be excused for thinking that the bright red bird with the blue nape and jet black bill was the male and the all-green bird with the yellow bill was the female. In this parrot, the typical colour scheme is reversed and it is the female who is brightly coloured. Females compete intensely for tree hollows in large trees that are suitable for breeding, but these are rare, with only one good nest site per square kilometre of rainforest. Males are a relatively inconspicuous iridescent green and cannot defend territories or mates; instead, they congregate at the nest hollows where they feed females and compete for access to them.

Competition among females is fierce because nest sites that remain dry during downpours increase breeding success. Females with drier hollows produce more offspring and attract more males for food provisioning. Females guard their nest sites for most of the year. A female has been known to commit infanticide and kill the resident's eggs or young in an effort to take over a nest site, and even to kill another female during a fight.

Before nesting, females display in the high branches of their nest tree. Males and competing females flying toward the nest tree generally get their first view of the territory owner from above and females are most conspicuous from this angle against the green background of the canopy. Once the eggs are laid, females remain in their nest hole but still display their bright red head from the nest entrance when they look out. The green colour of males is inconspicuous against a green background, but does contrast with tree trunks. Male–male displays and fights occur most often at the nest hollow, where males perch and knock each other off. While foraging for fruit, however, the male's fighting colours are effectively turned off because he spends most of his time in foliage.

Though people quickly appreciate the intensity of aggression in animals equipped with obvious weapons—lions, wolves, snakes, sharks—humans are mostly oblivious to aggression in birds because they pose no threat to us. Alfred Hitchcock's frightening but unrealistic movie *The Birds* was perhaps the only attempt to put birds on a par with truly dangerous beasts. Nonetheless, fighting and intimidation dominate the social lives of birds as individuals compete for food and mates, and this intense competition shapes their behaviour, colouration, and songs.

9 BIRD CITIES
Why Birds Live in Groups

Social interactions reach their zenith in bird colonies, where dozens, hundreds, or even hundreds of thousands of pairs breed side by side, not unlike in our large metropolitan cities. Colonies are noisy, chaotic places where individual territories are not much larger than the bird itself. There are many potential costs of living in huge groups, including intense competition for nest sites and mates, food depletion near the colony, attraction of predators to the buffet, and the spread of disease and parasites.

In the 1980s a hot topic in behavioural ecology was the question of why birds form such large, crowded colonies, and that question is still the subject of much debate today. Safety in numbers was the traditional explanation, because hundreds of pairs of eyes means that predators cannot sneak into the colony and, if they do, the huge number of nests means that the odds of any particular nest being attacked is low. For seabirds, colonies are often on offshore islands that many predators cannot reach.

The advantage of many eyes to defend against predators is clear in the lesser kestrel. These small European falcons nest in holes and prey mostly on large insects like grasshoppers and beetles. From 1993 to 2000, researchers studied lesser kestrels in northeastern Spain where they bred under tiled roofs of abandoned farmhouses surrounded by fields of cereal crops. Nesting sites were abundant and buildings attracted one or several pairs as well as colonies of some forty pairs. Nest predation determined who produced successful nests, or not, and nests in small colonies were three to five times as likely to be attacked by predators. Survival of breeding adults was also 10 percent higher in large colonies. Larger colonies tended to be in predator-free buildings, where rats were scarce and foxes could not reach the nests. Not surprisingly, individuals who switched colonies usually moved to a larger one.

This raises the question of why any pairs would nest alone or in small colonies in the first place. There may be costs of living in large colonies that are born more heavily by some individuals than others. For instance, young birds might preferentially form small colonies to avoid intense competition and then trade up as they get older. Alternatively, small colonies do sometimes enjoy good nesting success, and large colonies can be wiped out en masse by rogue predators that are attracted by the obvious bird activity. More discreet nesting sites might allow a bird to slip by unnoticed.

When I began my PhD at Yale University in 1987 I was swept up in the coloniality question, probably because my supervisor had devoted his career to the subject, and I decided to tackle a new angle. At that time, DNA paternity testing was in its infancy and I wanted to test the idea that the risk of being cuckolded was a big cost of living in large groups. With so many females packed into a small space, surely the opportunities for both sexes to sneak copulations were almost endless.

I decided to study the bank swallow, a small bird that digs 1-metre-long tunnels into the steep banks of rivers. During nest excavation, the female clings to a small depression on the bank and uses her beak and feet to fling the sand aside. Once a hole is started, she enters the partial tunnel and kicks the sand out of the tunnel, rather like a dog digging a hole. Colonies can be made up of thousands of holes peppering the upper portion of the steep incline.

My study site consisted of a dozen gravel pits in central Iowa; bank swallows are not particular about whether or not there is a picturesque river flowing beneath the colony. This region was rich with glacial deposits of sand and gravel, and bank swallows occupied many of the gravel pits. My job was to catch as many pairs as possible to get DNA samples, check their nests twice a week, and get DNA samples from nestlings. I sat for hours watching the entrance of marked pairs to see how often the female arrived or left unescorted by the male. In the background were the sounds of bulldozers and trucks working the other sections of the gravel pit.

Watching a colony is dizzying, as individuals and pairs come and go constantly, sometimes in waves. Birds shoot out of their tunnels like little brown missiles, the male often tailing the female closely. Dozens of holes have energetic puffs of sand flying out simultaneously, and arriving swallows disappear into the darkness carrying little pieces of grass for their nests at the far end of the tunnel. The scene is hectic and the individual birds take on a veil of anonymity—dozens of identical-looking brown birds against a greyish-brown backdrop.

I caught bank swallows by the dozens in mist nets stretched in front of the tunnels; this worked particularly well at dawn when I could catch the swallows coming out of their tunnel. A small army of field assistants helped pluck the birds out of the nets, and each bird received paint marks on its wings, so I could keep track of which pair was which, and a numbered leg band. To check nests, I built a "riparia-scope," named after the scientific name of this bird, *riparia,* that reflects its association with rivers. The scope was a long tube, with mirrors and a light at the far end, and a couple of lenses in the tube to magnify the image. For each tunnel, I gently pushed the scope all the way in, put my eye to the lens at the bank face, and peered at the hazy, dimly lit image to count the number of eggs and see if they had hatched. I had hundreds of nests to check, and spent most afternoons scrambling on slippery gravel slopes trying to see into nests. I came home hot, dusty, sun-baked, and with seemingly pounds of gravel and sand in my shoes.

The DNA part of my study did not work, as it turns out. I used my one and only research grant to pay a lab at Cornell University to test the blood samples. At that time, paternity testing was done by comparing the subtle genetic variants of enzymes that exist in a population. Though the enzyme does the same job for every-one, slight genetic mutations occur in some individuals and can be used to tell if a particular male is actually the father of the young in his nest. This is somewhat like using ABO blood types to compare parents and offspring. I received the bad news a few months after my field season ended; though two dozen differ-ent enzymes had been tested, most had only one or two variants, which was not enough to say much about paternity. This was before the days of DNA fingerprinting, which became possible just a few years later, and I decided to abandon the project. The next summer I was standing on a lush lawn on the shores of Lake Texoma, on the Oklahoma border, watching male purple mar-tins fighting over nest boxes.

I had the right idea with my original PhD thesis. Subsequent studies have confirmed that cuckoldry is common in large bank swallow colonies. In many species, sexual competition for mates increases with colony size. Cliff swallows nest in colonies of up to three thousand pairs, and attach their gourd-shaped mud nests not just to cliffs but to concrete walls, culverts, and bridges. Extra-pair copulations are observed more often in large colonies, and males from large colonies have super-sized testes and higher levels of testosterone in their blood streams. Large testes allow a sufficient sperm supply for a male to copulate frequently with his own mate and other willing females.

❧

Living in crowded conditions comes with a near-universal cost: increased disease transmission. Purple martin colonies are often overrun with nest parasites like feather lice and mites because so many birds live in close proximity. When we open doors to check nesting compartments and remove nestlings for banding, dozens of tiny mites crawl up our arms, giving us a taste of what it might be like to live there for a full month. Fortunately for us, they have a taste only for bird blood! The nestlings themselves are often seething with the tiny black specks, and as if that were not bad enough, they may also have large fly maggots attached to their toes and bellies.

Cliff swallows are plagued by swallow bugs, long-lived parasitic insects somewhat similar to bedbugs, which live year-round in the birds' nests and feed mainly on the helpless nestlings. Even a cold Nebraska winter is not enough to clean the nest, as when adults return to the colony in spring the female swallow bugs are waiting to jump on board and get their first blood meal, which will soon be converted to eggs. There are typically two hundred to five hundred swallow bugs per nest, all of whom will get their sole nutrition from the swallow family who lives there. Not surprisingly, the nestlings living in heavily infested nests are much smaller, and in some cases nestlings jump out of the nest to escape the bugs and are doomed because they cannot survive on the ground. The severity of swallow bug infestations increases with colony size, because the closely packed nests and abundant hosts are a perfect world from a parasite's point of view.

To find out exactly what impact the parasites had on the cliff swallows, my supervisor, Charles Brown, removed the parasites from half of the nests in a colony. Nests were fumigated every few days with a mild insecticide, and a thick sticky gel was painted on the concrete ceiling and wall of the culverts to prevent swallow bugs from crawling into the fumigation area. At the end of the breeding season, fumigated nests contained only twenty bugs per nest compared with an uncomfortable seven hundred bugs per nest for the non-fumigated side of the colony.

Since the swallows from fumigated and non-fumigated nests lived in the same colony, faced the same threat from predators, and had the same access to food, this experiment zeroed in on the real costs of parasitism. At colonies with more than three hundred pairs, those with infested nests produced half the number of nestlings compared with pairs from fumigated nests, and infested ten-day-old nestlings were 10 percent smaller in size. Given that swallow bugs can reduce nesting success by at least 50 percent, they are a major cost of living in a large colony.

Not only do swallows choosing a large colony face having their nestlings suffer from swallow bug attacks, but the swallow bug is also the middleman that transmits a virus that infects cliff swallows. The virus has a strangely ironic name, the Buggy Creek virus, named after the location where it was first found. The percentage of nests in Nebraska containing bugs testing positive for Buggy Creek virus increased with colony size, and in some larger colonies over 50 percent of swallows were at risk of exposure in their own nests. The effects of the virus on the health and survival of adult swallows are still uncertain.

In spring a first-time breeder knows if it is joining a large or small colony because the mud nests remain glued to cliffs and walls over the winter. Yet, thousands of individuals willingly choose parasite-infested ones, so there must be some advantage of living in large colonies.

Though my bank swallow study flopped with respect to paternity analyses, I was able to learn something interesting about the potential advantages of coloniality. Why would a Canadian student doing a PhD in Connecticut decide to do fieldwork in Iowa? Yes, bank swallows were common in Iowa, but the species is widespread and I could have worked in any number of states. I chose Iowa for two reasons; first, because the University of Iowa offered me free housing at their field station plus a thousand dollars for research expenses; and second, because Iowa is flat.

I needed flat open ground around the colony so I could watch the bank swallows as they flew off to feed over the nearby fields.

Charles had recently published a now-classic paper showing that cliff swallow colonies in Nebraska were important for exchanging information about where food can be found. When feeding nestlings, a cliff swallow that comes to the nest without food waits there until a neighbour arrives with a mouthful of insects, then follows the successful bird back to wherever it had found such a great meal. Insect swarms are unpredictable in space and time, and depend on temperature and wind direction and the behaviour of cattle and farmers; swallows follow moving animals and tractors that stir up insects in their wake. From one hour to the next a parent doesn't necessarily know where to find food and copying successful neighbours is faster than searching alone. One of the main advantages of colonial living, then, is being able to find food quickly for nestlings by using the colony as an information centre.

To test whether bank swallow colonies also functioned as information centres, I carefully watched the swallows coming and going from the gravel pits. I was surprised that the birds who fed nestlings were rarely followed away from the colony, and when birds did leave the colony at the same time, their flight paths split up before they started feeding. Despite many hours of watching, at several different colonies, there was no sign of birds following each other for food.

I wondered if this could be a quirk of my study area—that the food supply was not patchy enough to make it hard to find flying insects. A good test would be to watch cliff swallows, since I knew that in Nebraska they *did* follow each other. I took a few days off from my gravel pits and instead drove the highways looking for bridges with cliff swallow colonies. Watching from underneath, I found that even cliff swallows in Iowa were not following each other to feeding areas.

What was a huge benefit to swallows in one region was apparently unimportant in another, yet cliff swallows and bank swallows still formed large colonies in Iowa. My small study reflected what was being found on a larger scale—the costs and benefits of coloniality varied greatly among species and even between different populations of the same species. The traditional approach of studying coloniality, weighing costs and benefits, did not always show the balance strongly tipped toward the benefit side.

~~~

In recent years, a new way of viewing coloniality has emerged. The original cost–benefit approach assumed that individuals benefit by joining a group and thus colonies were viewed as serving a function such as reducing predation or enhancing food finding. The new approach considers bird colonies to be a by-product of birds using the same "rules" for habitat and mate choice. When numerous individuals choose the same location because food is abundant or predators are scarce, the result is a colony. Communal defence and group foraging do not cause group living, but rather arise after group living has evolved for other reasons.

Birds are nosy neighbours and constantly monitor the behaviour of competitors to know where to find food, where the best places are to breed, and even which individuals are the highest quality parents and future mates. Females can obtain information about mate quality by eavesdropping on male–male interactions. Female chickadees listen to dawn song contests between neighbouring males to learn which is dominant and the next day intrude onto the territory of the winner to mate with him. A bird can see how often a flock member captures prey, or monitor food deliveries to nearby nests, and thereby find rich food sites without searching single-handedly. This kind of public information is provided inadvertently, unlike the social signals used deliberately to intimidate rivals and impress mates. Young birds that are not old enough to breed spend much time visiting active colonies to gain information on how many pairs have young; this helps the youngsters select a good colony site for their future breeding efforts. The benefits of using public information are that it reduces the costs that come with solitary trial-and-error learning and often leads to more accurate assessment of resource availability and quality.

For humans, a full parking lot outside a restaurant is a sign that the food is pretty good, an empty concert hall is probably an indication that the performance will be a disappointment, and a daycare with bubbling, smiling, finger-paint-spattered children is one most parents would be comfortable trying for the first time. Seeing red flashing lights in your rearview mirror is a distinct possibility if you are driving too fast down a deserted highway, but when following a group of other cars you see the brake lights ahead and can slow down before you reach the speed trap. I often inadvertently information-share when walking around campus; when I stop to look up I am quickly copied by strangers who glance up, or even stop and stare. More than likely, I'll be admiring a red-tailed hawk flying overhead or watching a flock of blue jays whiz past carrying acorns—not useful information to undergraduate students but it demonstrates our innate tendency to watch neighbours.

Public information is used by black-legged kittiwakes when choosing a colony. These long-lived seabirds nest on cliffs, and one population in France has been studied for over fifteen years. Reproductive success on a given cliff is usually similar two years in a row, but over the longer run good nesting cliffs can lose their advantage due to outbreaks of blood-feeding ticks. Young kitti-wakes do not breed until at least three years of age and in that time gather information on nesting success late in the summer when they make visits to local colonies. Research showed that they joined colonies that did well the year before, using informa-tion that would not have been possible to gather in spring when breeding started. Colonies with high reproductive success one year grew rapidly the next year because of recruitment by first-time breeders and failed breeders that switched colonies.

The black kite is a loosely colonial raptor that, in the Italian Alps, nests on cliffs facing large lakes where it forages for patchily distributed fish. Nesting success is tightly tied to food availabil-ity and risk of predation by eagle owls, which create significant differences in habitat quality between colonies. Though nesting success is fairly predictable from one year to the next, habitat quality cannot be assessed solely on the basis of the cliffs' physi-cal features.

Young birds solve this problem by settling near other pairs that did well the year before, and then watching the experienced birds foraging to learn where to hunt. Young kites set up terri-tories close to pairs that produced an above-average number of young the previous year, suggesting that they had been monitor-ing the neighbourhood for a long time. Very high productivity of a territory in one year was followed by settlement of a new pair nearby the following year. This meant trouble for the original pair because of the amount of time spent fighting and trying to chase away the new neighbours, leading to low reproductive success despite the good food supply and low risk of predation. The established pair won direct confrontations with their new neighbours, but sheer persistence won out, for the young birds had little to lose and much to gain in the long run.

Many seabirds nest in colonies, are long-lived, and return faithfully each breeding season to the same nest site in the same colony. The advantages of using public information in dispersal and settlement decisions means that new unoccupied habitat is a blank slate, and thus individuals are reluctant to form new colonies from scratch. The adaptation that led to, and sustains, coloniality becomes a headache for wildlife managers trying to create new habitat or re-establish colonies in locations where they have been wiped out. No matter how much time, money, and effort is expended on habitat restoration and creation, many birds will not occupy an area unless there are already birds there.

The "social attraction" technique is used to trick prospecting birds into thinking habitat is occupied. Vacant habitat is equipped with decoys and playbacks in the hopes that some birds will stay long enough to attract additional recruits. If a critical mass is obtained, then nesting activity can be stimulated. Even if the colony is tiny at first, with a bit of luck and decent nesting success the first year, those founding breeders will return the next season and effectively lure other breeders to the site. This artificial seeding of colonies has worked to establish colonies of Atlantic puffins on islands off the coast of Maine. Bird lures included mirrors and unrealistic two-dimensional cutouts of puffins; though birds can obviously tell the lures are not real, the stimuli are enough to encourage curious birds to at least visit the site and linger long enough to encounter a live bird.

Common murres occupy the same colony sites for decades, new colonies are rarely formed, and abandoned colonies are rarely recolonized. Common murre colonies in central California declined suddenly by over 50 percent from 1980 to 1986. One colony, Devil's Slide Rock, had two thousand murres in the early 1980s but by 1986 the colony was devoid of breeding murres. The population crash was a result of high adult mortality caused by two man-made problems: an intensive gill-net fishery, and several oil spills, particularly the 1986 *Apex Houston* spill. More than six thousand common murres were found dead in just two months following this spill. For the next decade, the Devil's Slide Rock site had no breeding murres, though occasional visitors were seen.

To re-establish the murres at Devil's Slide Rock, an artificial colony was created in January 1996 by adding hundreds of life-sized murre decoys, an audio system that broadcast murre calls, and mirrors so that reflections of live birds would simulate movement and behaviour. In early June, the time when central California murre colonies normally would be nesting, dozens of wooden egg and chick decoys were added among the adult models to simulate successful breeding activity. All the decoys, including the eggs, were hand-painted to look realistic.

That first year, about fifteen murres were seen daily at the colony site throughout the breeding season. The vast majority of sightings were of visitors near a decoy, and not in the many suitable nesting areas without decoys, suggesting that social attraction was working. Murres seemed to congregate most often near mirrors, and four of the six pairs that actually nested in 1996 settled within 60 centimetres of mirrors. Within two years there were thirteen breeding pairs, and by 2004 the colony was solidly established with almost two hundred breeding pairs.

Social attraction has also been used to restore historical breeding colonies of common, arctic, and roseate terns along the coasts of Massachusetts, New Hampshire, Maine, and New Brunswick, and Caspian tern colonies have been established on artificial islands in Lake Ontario. Biologists used the same tool kit of decoys and audio playbacks of vocalizations to establish and maintain colonies at sites where tern nesting had never before been recorded or had not occurred for decades.

Rice Island, in the Columbia River estuary in Oregon, was home to the largest colony of Caspian terns in North America, perhaps even the world. The island is artificial, made from dredgings to open the river to more boat traffic, and is just over 30 kilometres from where the river empties into the Pacific Ocean. So why did biologists from the Oregon Fish and Wildlife Service want to move these birds off the island?

Caspian terns are the world's largest tern and sport a massive orange-red bill for catching fish. They fly low over the water looking down, stop and hover when they see a fish, then make a spectacular head-first plunge under water to catch the fish. Studies had shown that about 80 percent of their diet during the breeding season was juvenile salmon. Columbia Basin salmon stocks are listed as endangered under the U.S. Endangered Species Act, putting biologists in a conservation and political bind. Salmon fishery managers were understandably outraged that Caspian terns on Rice Island were eating millions of young salmon each year.

The solution? Move the precious tern colony to a site where they would eat less salmon. The best prospect was East Sand Island, some 25 kilometres downstream and near the Pacific Ocean, where biologists hoped the terns would feed instead on marine fish. Caspian terns had nested on East Sand Island between 1984 and 1986, but not since. Other seabirds on East Sand Island fed mainly on marine fish, giving biologists hope that Caspian terms would do the same.

East Sand Island had been used as a dumping ground for dredgings, so the first step was to bulldoze the mess and create decent nesting places with smooth sand. Next, 380 Caspian tern decoys were commissioned and set in place with four audio-playback systems that broadcast vocalization from terns at the Rice Island colony. Next, a few glaucous-winged gulls (nasty nest predators) were humanely removed from the site intended for terns so that prospectors would take a second look. Finally biologists provided a bit of discouragement for the Rice Island colony. Caspian terns prefer open patches of bare sand for nesting, so winter wheat was planted in some parts of the traditional colony to encourage dispersal.

What a stunning success story! All Caspian tern breeders shifted from their salmon-eating habits on Rice Island to their new home on East Sand Island. In 1998, over eight thousand terns were nesting on Rice Island and none at the new site; three years later this was reversed and Rice Island no longer had Caspian terns. The biologist's gamble had paid off, and as a bonus, the terns nesting at their new home on East Sand Island exhibited much higher nesting success.

The scientific article in which this research is published presents something of an understatement: "Caspian terns that nested on Rice Island were successfully relocated to newly restored habitat on East Sand Island over a 3-year period" and "The short-term advantages to both juvenile salmon and Caspian terns associated with the relocation of nesting terns from Rice Island to East Sand Island are evident." This single population of Caspian terns represents 20 percent of the North American population and 10 percent of the global population. Creating new colonies is important; with so many eggs in one basket, a disaster such as a disease, a storm, a predator, or an oil spill at this one site could have a major impact on this colonial bird.

Breeding in colonies, by definition, concentrates the population and can make it vulnerable to disturbance, habitat destruction, and disease outbreaks. American white pelicans are colonial and almost half the world population breeds at four sites in the northern plains. Sustained productivity at these colonies is crucial to the health of the entire species, but in 2002 nesting success dropped dramatically. The culprit was West Nile Virus (WNV), introduced to North America in 1999 where the first sign of a problem was dead crows littering lawns in New York City. Before WNV arrived in the northern plains in 2002, chick mortality was less than 4 percent, but it jumped to as high as 44 percent in the following years. Diseased pelican chicks first showed signs of illness when they had trouble holding their heads up and maintaining balance, and within days were unable to move. By 2007, all three colonies were suffering high chick mortality. Pelicans are extremely long-lived, so they can ride out a few years of exceptionally bad nesting success. Their low reproductive rate, however, also means that genetic resistance to WNV is likely to spread far too slowly.

Seabird colonies on islands off the west coast of North America have plummeted in numbers in recent decades. The food supply for seabirds in the eastern North Pacific comes from an upwelling-dominated current called the California Current System that stretches from British Columbia, Canada, to Baja California, Mexico. This current can be extremely productive in some years but is highly variable due to the natural climatic cycles of El Niño driven by changes in surface ocean temperatures. In the past thirty years, there have been warmer and more frequent El Niño years as a result of human-caused climate change. With warmer waters, the biomass of zooplankton has crashed, causing a chain reaction affecting the abundance of fish and seabirds.

The Cassin's auklet eats krill, a type of zooplankton, and adults spend the day at sea filling up their crops and return at dusk to feed their waiting chick. The colony at southern California's remote Farallon Islands declined by roughly 75 percent from 1971 to 2002. In 2005 and 2006 few of the eggs in the colony of some twenty thousand pairs even hatched. During the 1992–93 and 1997–98 severe El Niño years, annual survival of adults dropped from 80 percent to as low as 40 percent and about half the adults did not even attempt to breed.

Other species up and down the coast have suffered a similar fate: tufted puffins in Oregon, rhinoceros auklets in California, and common murres in British Columbia. These seabirds feed their young on fish, and when ocean temperatures are warm, nesting success is almost nil. As currents and upwellings shift, parents have to go farther from the nesting colony to find food and cannot return to the colony daily, which leads to certain death for their eggs or young.

The problem of ocean warming threatens colonial seabirds worldwide. King penguin parents do not have time to complete a full reproductive attempt, from courtship to fledging young, in a single short summer. This means that birds are still breeding when their marine food supply is at its annual low in winter, and even small alterations due to climate change could devastate the population.

Researchers studied the survival of king penguins on Possession Island, which is in the south Indian Ocean, about halfway between Madagascar and the Antarctic. This out-of-the-way island is home to one-third of the world's king penguins. Small electronic tags were implanted under the skin of hundreds of penguins, and receivers were buried along the traditional path that the penguins use to march from the inland colony to the

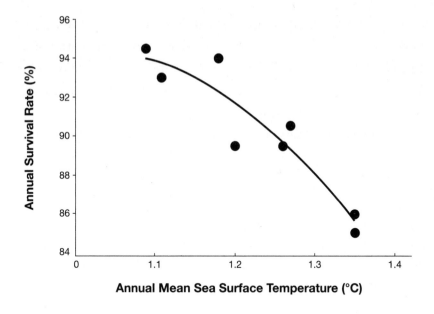

Figure 9.1. *The annual survival rate of king penguins on Possession Island, in the southern Indian Ocean, decreased dramatically when sea surface temperature was warm in preceding years. (After Le Bohec et al., 2008.)*

sea. This allowed researchers to automatically receive data on penguin movements, and to know which adults returned to the colony over a nine-year period. The probability of surviving to the next year dropped sharply with increasing sea surface temperatures in the preceding years (Figure 9.1). There was a two-year time lag between the low food supply in warm years and the lower survival rate of adult penguins. A warm summer leads to poor sea ice during winter, which in turn reduces the amount of krill available to foraging penguins. The time lag could be a result of birds having to work harder the first year to find food.

Penguins experienced an immediate decline in breeding success during summers with warm oceans, because phytoplankton productivity at the bottom of the marine food chain dropped, causing a fall in their summer prey (larval fish) and winter prey (small squid) that are fed to chicks. With food more scarce, parents cannot easily refuel after an incubation bout and may abandon the nest, and they are away from a chick longer because they have to travel farther from the colony to find food.

The Intergovernmental Panel on Climate Change forecasts a warming of 0.2 degrees Celsius per decade for at least the next two decades, and there is little that seabirds can do behaviourally to adjust to these new conditions. Colonial seabirds are at the mercy of their food supply and our ability to bring climate change under control. Seabird colonies are highly sensitive to food supply, which in turn depends on the ocean currents, which change with ocean temperatures.

10 FREQUENT FLIERS
The Demands of Migration

A bird has only so many hours in a day, and faces trade-offs between investing time and energy in one activity versus another. A female who lays extra eggs in her first nest may be able to raise a super-sized brood of young, but there is a future price to be paid for her ambitions. She may produce smaller and less successful offspring, take longer to begin her second nesting attempt, and be in poorer condition herself at the end of summer. Many studies have shown that high reproductive effort in one year can even lead to lower survival and breeding success the next year. This tug-of-war between reproductive success and a bird's health and survival is seen most vividly in migratory songbirds.

One might think that a female should go all out during the short breeding season if the chances of her surviving to breed again the next summer are rather slim. In hooded warblers, as with most migratory songbirds, only about half the adults survive their migratory journey and return to breed. Nevertheless, many females who successfully produce young from their first nest do not go ahead and attempt to breed a second time that summer. This seems odd because these "double-brooded" females produce an average of two extra offspring, which amounts to a 60 percent increase in annual productivity. Double-brooded females are not older, do not live on better territories, and do not get more help from their males when feeding the first batch of young. One of my first graduate students, Lesley Evans Ogden, set out to discover why so many females were holding back.

The nesting season for hooded warblers runs from mid-May until mid-August, and each nesting attempt takes about thirty days from the start of nest building until the young leave the nest. The parents' job is not over even then, because the young birds must be fed for almost three weeks more before they can care for themselves. One successful nest takes about half the breeding season. A female only has time to double brood if her mate takes sole care of the fledglings while she lays and incubates the next clutch of eggs, saving her about three weeks. Lesley also found that females who had a one week head start on their first nests in May, because they arrived earlier in spring, were much more likely to try to rear two families.

Time is also tight at the end of the breeding season because adults must undergo their annual moult of feathers in August, prior to migration. During moult, the wing and tail feathers drop out in a strict and predictable pattern leaving the bird looking scruffy and less agile. In mid-August, dense tangles of raspberry and wild rose are often occupied by hidden birds whose location is given away by an occasional *chip* call. A hooded warbler lays low, skulking in the same patch of vegetation for several weeks during the heaviest phase of moult. Eventually, the old tail feathers fall out en masse leaving the poor bird looking like it had a close call with a cat.

Moulting is one of the most costly times of year for a bird. During heavy moult songbirds cannot fly well and so are more likely to be killed by predators, they have less insulation so have to put more energy into staying warm at night, and the large number of feathers that are grown in such a short time require extra nutrients and energy. Double brooding allows a pair to produce extra offspring, but breeding late in the summer means less time to recover from parenting before moult begins. In many cases, the bird runs out of time and has no choice but to feed young and moult at the same time.

The first birds to start moult, in mid-July, are usually single-brooded parents that finished caring for their young by the end of June. For pairs that double brooded, most did not begin moult until early August. Double-brooded birds did not catch up by growing their feathers faster, probably because this would take so much extra energy. Throughout August, double-brooded birds lagged three weeks behind on their moult schedule, and, in addition to being late, some still had the job of feeding fledglings. Females may face the greatest burden because most males reduce their feeding effort to second broods, leaving females to do most of the work. One male abandoned his new family altogether so he could start moulting his feathers four weeks ahead of his mate.

Multitasking by overlapping moult and parental care takes an immediate physical toll on the bird at a time it can least afford to be stressed out or late. August is also a time for birds to fatten up and get ready for the 3,000-kilometre flight to Mexico. The delay in moult may be especially costly for adults and their young if their migration and subsequent arrival on the wintering grounds is also delayed. Hooded warblers are territorial on their wintering grounds, so a late arrival may mean the bird gets a poor-quality territory in dry, scrubby habitat, or worse, no territory at all. The real cost of double brooding lies not only in the short-term energetic costs of doing two tough jobs at once but also in the consequences of this for the bird's migration to the wintering grounds.

Once a songbird leaves its breeding territory we know next to nothing about what happens during the next eight months as it travels to, and lives in, some tropical country before returning to breed. We do know that migration is dangerous, since only about half the breeding population returns after the round-trip journey. Until recently, there has been no way to track the movements of individual songbirds over such long distances.

～♪

Songbirds have one of the most spectacular migration movements in the world, travelling thousands of kilometres between the temperate zone and tropics each spring and fall. In North America, billions of songbirds pour south each fall in such large numbers that their migratory waves show up clearly on weather radar. Migration from Europe to Africa and Asia is equally impressive. Larger birds, like eagles and falcons, have been tracked using satellite transmitters, but these weigh far too much for a small songbird to carry. Our current understanding of individual songbird migration comes from brief snapshots of the journey via radar images, opportunistic recaptures of banded individuals, and studies of migrants on the ground refuelling. One exceptional study followed radio-tagged thrushes for one or more nights by airplane!

The Holy Grail for those of us studying migratory songbirds has been some device that allows us to track individuals over hundreds, or thousands of kilometres. One idea was to attach tiny cell-phone devices to the bird's back that would automatically signal cell-phone towers near the bird's flight path. Since it is getting harder and harder to find anywhere that is out of cell-phone reach, this approach would give a detailed and instant feedback on the bird's location. The drawback is its short battery life, only a few weeks, but at least part of the journey could be monitored.

Another idea, championed by Martin Wikelski, then a professor at Princeton University, was to put radio receivers in space, on either satellites or the international space station. Just as radio astronomers point antennas into space to locate faint radio sources against background noise tens of thousands of times stronger, Wikelski proposed, wildlife biologists could point antennas toward earth from the space station to locate small radio transmitters attached to birds. Despite much work on these ideas, Wikelski noted that the one six-day airplane tracking of a Swainson's thrush in 1973 by Bill Cochran, a pioneer of radio-tracking birds, remains our best global data set describing the individual movements of an estimated forty billion songbirds migrating annually among continents.

Imagine my shock, then, when I was standing in a hallway in Veracruz, Mexico, a few years ago at a large international ornithology conference. I had picked up a cold on the flight down and was not in the mood to sit in an air-conditioned lecture hall listening to my fiftieth talk. What I really wanted was a nice lunch at one of the outdoor patios overlooking the Gulf of Mexico, and I was killing time by wandering past the many scientific displays when I heard a voice with a thick Eastern European accent say, "Have you seen my poster?"

I turned around and saw a heavy-set middle-aged man with a thick grey beard looking at me eagerly. To be polite, I took a step closer to the poster on the wall. There was a picture of an albatross, a map showing how it had flown around the southern hemisphere a few times, and a hand delicately holding a "newly miniaturized 1.5-g geolocator" between thumb and forefinger. The man's name was Vsevolod Afanasyev, from the British Antarctic Survey, and I quizzed him for the next hour about using geolocators to track migratory songbirds (I missed lunch too).

These devices detect light levels twenty-four/seven, and have a clock too. Over weeks and months the raw data show the predictable up and down cycles of sunrise and sunset. Under perfect weather conditions, and assuming the geolocator is in the open and not shaded, you can determine the bird's geographical location because the timing of sunrise and sunset is known for everywhere on the planet, every day of the year. (Rainy and cloudy weather obscures sunrise and sunset, though, which lowers the field accuracy to about 150 kilometres.)

As soon as I got back to Toronto after the conference, I contacted Vsevolod and his colleague James Fox to order fifty new geolocators customized for songbirds. It's easy to mount tracking devices on a bird's back, as I've done with radio-transmitters for many years, but the feathers would cover the light detector built into the geolocator. James simply added a short stalk to their new geolocator, to raise the light detector up above the feathers.

You need a songbird large enough to carry a 1.5-gram load, and this was no problem for me because I already had students studying purple martins and wood thrush, both about 50 grams. Purple martins live in open areas, migrate in the daytime, fly to South America, and are highly gregarious year-round. Wood thrush are almost the opposite: they live in the forest, migrate at night, fly to Central America, and defend territories year-round. By using both species, we could compare the migration of birds with two very different lifestyles.

To get any information at all, we have to catch the bird when it comes home in spring to retrieve the geolocator and download the light data onto a computer for analysis. As the bird travels east or west, the actual time of sunrise logged on the geolocator will be earlier or later the next day. A martin roosting on the Gulf Coast of the Florida Panhandle during fall migration will see the sunrise long before a martin that roosted under the Lake Ponchartrain bridge in New Orleans. This east–west location is measured by longitude.

Latitude is a bit harder to estimate, and is based on day length. During northern hemisphere summers, days get longer as you go farther north. In winter, however, days get shorter as you go north. As a martin travels south in fall, the days get shorter until it reaches the equator, and then get longer if the bird continues into southern Brazil. Another glitch is the awkward times of year when day length is equal everywhere on the planet and latitude cannot be estimated. These are known as the fall (21 September) and spring (21 March) equinoxes, which most people mark on their calendars as simply an official change of season. The two weeks before and after an equinox is a "blackout" time for latitude, but we can still estimate east–west position and so have some idea of where the birds were.

Our collaboration was off and running, though I suspect James thought I was a bit of an idiot for wanting to put a light-detecting device on a bird that would spend most of the winter living in the dark understory of a rainforest. It was urgent that we try, however, because wood thrush populations have declined steadily in the past few decades. The evolutionary balancing act between reproductive effort and subsequent survival is a challenging one at the best of times, and the heightened threats that songbirds face on migration mean that time-tested strategies may no longer work in today's world.

Our first geolocators were put on purple martins, because these large swallows could carry the tags and I was already working closely with the Purple Martin Conservation Association in Pennsylvania. There is no doubt that purple martins are masters of the sky, plucking dragonflies out of the air with ease and then flying into the impossibly small hole of their nesting compartment at breakneck speed. People who put up nesting sites for martins spend many a lazy evening admiring the acrobatics of their tenants, who seem more at home in the air than they do on land. The martin's long tapered wings and streamlined body say it all—flying machine. In late summer when the nesting colony is strangely quiet, I often see martins flying overhead on migration and hear an occasional gurgling song. I can hardly believe these intrepid travellers are going all the way to Brazil.

We know that martins overwinter throughout much of South America, where they spend their days feeding over tropical rainforests, rivers, and farmland and at night gather in large roosts. Martin are known to roost in towns in the deforested areas of southern Brazil, in São Paulo province, and even on the pipes of an oil refinery in Manaus, on the Amazon River in northern Brazil. Though we knew a lot about each end of the journey, we knew next to nothing about how individual birds make such a marathon trip.

The details of the purple martin's amazing journey are no longer a secret to us thanks to geolocators. In the summer of 2007, during two "martin mornings," we put the tiny devices on purple martins nesting at the Purple Martin Conservation Association's main colony in Edinboro, Pennsylvania. The geolocators are carried like a backpack, with the loops going around the legs, not the wings, and weigh about the same as a dime. When we released our birds, they flew out over Edinboro Lake, gave themselves a few good shakes (sort of like a wet dog?), and then, as we watched, started feeding high over the lake. All returned to their nests and fed their young, and over the next two weeks gradually disappeared one by one from the colony after the young had fledged.

Then began the long ten-month wait to see if our martins could carry the geolocators the entire 13,000-kilometre round trip to Brazil. Birdwatchers eagerly look forward to their first migrants each spring, but we were on pins and needles waiting to see if Emily Pifer, the PMCA biologist, would spot one of our geolocator birds. The good news arrived on 30 April 2008 via an e-mail from Emily. "Hi Everyone. Just wanted you to know I was down at Indianhead this morning doing some band reading and looking for geolocators. I saw a geolocator on an ASY-F . . . she is Yellow 2551. I saw her sitting on the porch at WH-44 for awhile." Emily was looking at the first migratory songbird, anywhere in the world, for whom we would know its arrival time on the wintering grounds, where it had spent the winter, and how quickly it had come home. As a bonus, a few days later at the colony Emily saw a second female carrying a geolocator. The next year, we recovered another three priceless geolocators from our colony.

I had always thought of martins as lazy when it comes to migration, though of course the distance covered is impressive. Martins fly during the daytime and feed along the way, and they stop at night to gather in huge roosts. What's the hurry? There is plenty of food up for grabs in the warm southern states, and roost sites have been mapped out using weather radar showing lots of martin "motels" along the way. I am in good company, since the official species account for the purple martin in *The Birds of North America* says, "Probably follows typical swallow pattern of leisurely movement, with migration in both directions spanning several months."

To my astonishment our geolocator martins had wasted no time flying south to Mexico once they actually began migration. Our birds stayed in northern Pennsylvania until late August and then, without warning, flew to the Gulf Coast states, across the Gulf of Mexico, and arrived at the Yucatán Peninsula near the city of Mérida within five days of leaving Pennsylvania. The martins had somehow covered 2,400 kilometres, including an 800-kilometre over-water flight, in less than a week (Figure 10.1A).

Recoveries of the same banded bird hundreds of kilometres apart can help us connect the dots of migration movements. Though the PMCA has banded more than twenty thousand martins in northwestern Pennsylvania, very few of these banded birds have ever been seen or recovered in Central or South America. The first was a nestling banded by the PMCA on 23 July 2004 and found dead in October that fall by Guillermo Castillo Vela at San Felipe, on the northern coast of the Yucatán Peninsula, Mexico. This single band recovery was suggestive of a Gulf of Mexico crossing, and now we know that all of our Pennsylvania martins make this marathon leg of the trip.

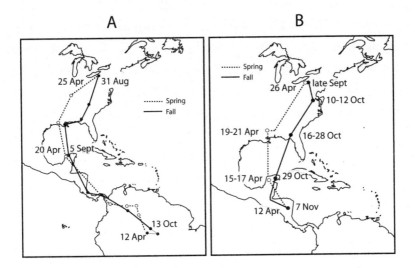

Figure 10.1. *This figure shows migration routes and timing for (A) a purple martin and (B) a wood thrush. The two were tracked using light-level geolocators. Positions are accurate to about 150 kilometres. (After Stutchbury et al., 2009.)*

Yellow 2551 arrived in Brazil in early October and stayed in the region near Manaus, Brazil, where martins are known to roost on the pipes in the oil refinery. The noisy maze of pipes reeking of petroleum is across the river from lush rainforest. I don't think our martin was actually roosting in the refinery, however, because her geolocator showed no signs of artificial lighting at night. All five of our martins overwintered in the Amazon rainforest, most near Manaus.

Yellow 2551 spent her last night in Brazil on 12 April, and then rocketed northward to arrive at her breeding colony by the night of 25 April. She flew about 7,000 kilometres in only thirteen days (over 500 kilometres per day)! For our five martins, the average time to complete spring migration is only twenty-three days (300 kilometres per day).

How does this flight performance compare to that of other migratory songbirds? Spring migration speed of five species of European warblers banded and re-caught in Britain and Europe was as fast as 225 kilometres per day. In North America, the radio-tagged thrushes followed by Bill Cochran in his airplane flew at an overall migration rate of 112 kilometres per night during spring migration. Even our slow-poke martin, who went "only" 220 kilometres per day during spring migration, flew much faster than expected. Spring migration is faster than fall migration, probably because birds are in a hurry to claim breeding territories and mates.

Seeing the migration maps for these martins raises intriguing questions. Do males, who are slightly larger, migrate even faster than females? Do older birds migrate faster than birds migrating for the first time, and do they overwinter farther north so they can get home more quickly in spring to claim their nest sites? Under what circumstances do martins fly overland around the Gulf of Mexico instead of making the risky over-water crossing? Do all our martins in northwestern Pennsylvania spend the winter in northern Brazil, flying over rainforests every day?

Tracking individuals to their wintering areas is important for understanding what threats martins face when away from their well-cared-for nesting apartments. Martins that feed over agricultural fields in southern Brazil are undoubtedly exposed to dangerous pesticides that are used in large quantities throughout most of Latin America. Perhaps overwinter survival is highest for birds wintering in the forested regions of northern Brazil. There is great concern that climate change will hurt our migratory songbirds, but forecasting these effects is difficult because weather conditions vary so much over the winter range of migrants. Northeastern South America is expected to get hotter and drier, but southern Brazil is expected to become wetter. Since martins depend on flying insects for their food supply, overwinter survival may differ for purple martins living in the northern versus southern parts of their winter range.

~~~

We also put geolocators on wood thrush, because this species is declining rapidly and is vulnerable to deforestation on its breeding and wintering grounds. In two years, we retrieved geolocators from fourteen wood thrush after migration. Fall migration was relatively slow but variable, with individuals arriving on their winter territory between mid-October and early December (Figure 10.1B). Like the martins, wood thrush crossed the Gulf of Mexico (especially on their way north) and spring migration was quite rapid. Most wood thrush flew from Central America to northern Pennsylvania (3,600 kilometres) in only two weeks. One female took the scenic route and flew overland around the Gulf of Mexico, flying an extra 1,000 kilometres and arriving at her breeding territory weeks behind everyone else. I couldn't help wondering if this bird had spent the winter as a non-territorial floater, or had lived in dry, scrubby forest, and was in poor condition when it was time to start her journey.

As with the warblers, many female wood thrush suffer predator attacks on one nest after another, and finally give up on nesting in late July without having produced a single offspring. Other females are fortunate enough to escape the attention of predators and produce two or even three successful broods of offspring during the summer. This parental effort is admirable, but what price do these parents pay for their double duty? The high reproductive effort of double-brooding pairs is expected to reduce their energy stores and delay both their feather moult and their southward migration. My students Tyler Done and Elizabeth Gow used blood and feather samples to study the consequences of high reproductive effort in wood thrush at our study site in Pennsylvania. The idea was to repeat the study that Lesley had done on hooded warblers many years before, but this time using high-tech blood testing, radio-tracking, and geolocators.

A picture tells a thousand words, but a drop of blood or a feather can tell an ornithologist far more. We have already seen how blood samples are used for paternity analyses in birds, revealing high levels of cuckoldry in many species. Just as for humans, a blood sample can also be used to reveal the physical condition and health of a bird through the hormones and metabolites that circulate in its tissues. Birds that are barely keeping up with the energy demands and stress of breeding will pay a physiological cost that can have lasting effects long after the breeding season.

Tyler acted as the bird doctor and was in charge of getting blood samples and doing lab tests. In this case, it was the doctor rather than the patient who spent hours waiting for his "appointment." Tyler spent much of his summer sitting in a small field chair, alone in the woods, watching for hours on end and waiting for his thrush to blunder into a mist net. It was critical to take the blood sample within three minutes of catching the bird, because the very act of being captured alters the blood chemistry. Wood thrush are relatively large songbirds, almost as big as robins, so the few drops of blood Tyler took were not missed. One of the important hormones we measured was corticosterone, a stress hormone, which increases several minutes after the bird experiences a stressful event. Since Tyler's quarry could hit the net at any moment, a lapse into daydreaming or nodding off could mean missing the capture event and having to start all over again the next day.

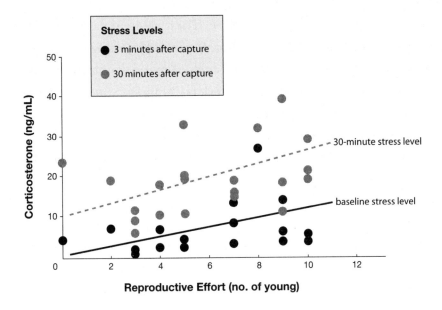

Figure 10.2. *This figure shows the concentration of corticosterone, a stress hormone, in the blood plasma of wood thrush captured while moulting. The birds that had produced more young had higher stress levels. (Done et al, 2010.)*

The nets were placed about 20 metres from the nest, and Tyler could usually catch both parents in half a day. If he was unlucky, as it happened several times, he encountered a bird who was wary of the net and his vigil lasted for several days. When he finally had the satisfaction of seeing the target fly into the net, Tyler jumped to his feet, grabbed his little medical tray of needles, capillary tubes, and cotton swabs, and sprinted like a madman to the net. The precious few drops of blood were soon sitting safely in a cooler.

The stress of parenting was clearly seen in wood thrush captured during August, when birds had finished breeding and were undergoing their annual feather moult (Figure 10.2). First, we measured their corticosterone levels within three minutes of capture to establish a baseline. Then we held each bird in a bag for thirty minutes to test how it coped physiologically with a stressful event. Birds that had hatched many young during the nesting season were more stressed, both upon capture and after thirty minutes. Their corticosterone levels were double those of birds that had raised only a handful of nestlings.

Elizabeth is the feather expert and was in charge of catching the wood thrush in August, during the feather moult and pre-migration period, when they were no longer visiting their nests. Wood thrush are shy and secretive even during the peak of the breeding season, and become more so in August. The first year of this study, we optimistically thought we could catch them, through sheer determination, by blanketing the area with thirty to forty nets each morning and having a team of four people check the nets at regular intervals. This Herculean effort yielded only one adult wood thrush in the nets every two days.

As a backup plan, we had put radio-transmitters on birds in June and July when they were still nesting. These individuals could be caught in August only because we knew their where-abouts and could put up walls of nets and then herd the birds toward our waiting trap. This sounds easy if you don't mind wad-ing through dense tangles of razor-sharp wild rose and black-berry bushes that tower over your head. Thrushes, like warblers, are shy during moult and often hide out in the densest, thickest vegetation they can find. The radio signal tells us the bird is right in the middle of the mess, so we set up nets in the forest around the edge. Forming a reluctant line, with the person at either end holding a radio receiver and antenna, we march through the patch and flush the bird toward the nets. If our line is not tight enough, the bird stays put and gets behind us . . . so we reor-ganize and try again. After a few days of this, despite thick blue jeans and field shirts, our arms and legs were spotted with thorn punctures and scratches. Adding insult to injury, we accidentally caught a dozen hooded warblers every day, reminding us that other birds are so much easier to work with.

As expected, wood thrush that tried to raise a second family were late starting their moult, so much so that they began their migration still carrying tattered old feathers. Feathers contain almost as much information as blood, revealing the geographic location of the bird at the time they grew. The new material that is used to make the feather contains stable isotopes from the rainwater in that region. Stable hydrogen isotopes in rainfall change predictably as one goes from north to south, and these natural isotopes are incorporated into plants and so on up the food chain. Elizabeth tested the feathers of wood thrush who returned in spring to determine if, on fall migration, their feathers grew in Pennsylvania, on migration, or on their wintering grounds in Mexico. She found that many double-brooded wood thrush overlapped moult with migration, and some even finished their moult in Latin America.

What are the costs of being energetically stressed out in August during moult and of overlapping moult and migration? Our recent geolocator research clearly shows that birds in poor condition during moult have a slower migration and arrive later on their wintering grounds. Wood thrush are voracious fruit eaters, and some have a slow, leisurely fall migration, probably to take advantage of the large supply of fruit along the way. What we don't yet know is whether late moult, or starting the journey in poor condition, also causes birds that arrive late in Latin America to get poor-quality territories. One of my new students, Calandra Stanley, is working at La Selva Biological Station in Costa Rica to link feather moult to arrival times, stress levels, and winter territory quality.

Given the ever-dwindling amount of rainforest in many tropical regions, migratory birds may face increasing costs of putting too much time and energy into breeding. If wood thrush are evolutionarily hard-wired to start a second brood whenever time permits, this could drive birds into a catch-22 of moulting in poor condition, which ultimately reduces their chances of survival as they run the gauntlet of forest patches on their journey to Central America and back. Breeding behaviour and winter survival may be more closely linked than ever before.

~~~

Recent studies have shown that migration behaviour can respond to changes in the environment. Populations of blackcaps in southern Germany and Austria that historically migrated to Portugal for the winter have, over the last thirty years, shown an increasing tendency instead to spend the winter in Britain. A survey of sightings by British backyard birdwatchers showed that in the 1960s blackcaps were seen only occasionally during winter, but in recent years almost a third of backyard feeders were home to blackcaps. The first clue as to the origin of these British visitors came from occasional recaptures of birds that had been banded in Germany.

More complete mapping of migration came from analysis of stable isotopes in the birds' claws. Claws grow very slowly and once the tissue is formed it does not incorporate new material (like hair and feathers), so it reflects the bird's location a few months earlier. Researchers clipped the very tip of one claw of a sample of blackcaps returning to the breeding grounds in Germany, and found that about one-third of the population had recently been living in Britain.

This shift in migration direction, and distance, has a strong genetic basis. Researchers captured adults wintering in Britain, brought the birds back to aviaries in Germany, and then allowed the birds to pair up and breed. The young from parents who had used the new migration route were tested in early autumn to find out if they had inherited this behaviour from their parents. Although natural migration unfolds over thousands of kilometres, the behaviour can be studied in a cage. Individuals are put in a small, funnel-shaped cage with a wire roof, which is positioned where the bird can see the night sky. Many songbirds use the stars to tell north from south, and after sunset the direction in which the caged bird is attempting to travel, and its overall activity, are a measure of its true migration instinct. Young birds with British parents showed a clear tendency to migrate west and northwest (that is, toward Britain, in real life), but most young taken from wild nests in the German population tried to go southwest as if their destination was Portugal.

The use of the new migration route has increased rapidly in just a few decades because of recent evolution; individuals who inherit this gene from their parents subsequently have relatively high survival and reproductive success, so the gene becomes more common each generation. The high number of bird feeders in Britain increases the winter survival of adults and may have made this new migration route possible by allowing those wandering individuals to return to breed.

British birds also get a head start on competition for breeding territories. The blackcaps who travelled to Britain began spring migration earlier than Iberian birds and had a shorter distance to cover, hence they arrived at the nesting areas earlier. Early arriving males claimed the best territories and early arriving females had high reproductive success by virtue of pairing with the best males on the best territories. Even in the wilds of Germany, birds tended to pair up according to where they spent the winter, an indirect result of arrival time. This further increases the spread of the favourable gene because the British pairs produce young who also go to Britain, who in turn will produce grandchildren who have the same migration behaviour and advantages.

Though the tendency to head west versus southwest is genetically based, the earlier spring migration of birds wintering in Britain may involve little or no genetic change. The annual stages of a bird's body (reproduction, feather moult, migration) are determined by changes in day length, which in turn trigger changes in the production of hormones that control the bird's physiology. As the winter nears an end, the very rapid change in day length in Britain compared with more southerly latitudes triggers a faster change in hormones that stimulate growth of the ovaries and testes. Blackcaps overwintering in Britain respond more quickly to the increasing day lengths of spring and thus are ready to breed about twenty days sooner than their southern counterparts.

This recent change in migration behaviour appears to be driven by climate change. Decades ago, the occasional birds that did overwinter in Britain likely arrived on the breeding grounds too soon, when weather conditions could be harsh and food supplies low. With warmer springs, birds that arrive earlier on the breeding grounds do not pay a heavy price, and in fact out-produce blackcaps who overwinter at southerly latitudes. One might expect that in fifty years or so, the traditional migration route to Portugal will be rarely used or disappear altogether.

European blackbird migration too has been dramatically altered by climate change, though in this case the effect is highly local and occurs as a result of living in cities. As recently as 150 years ago, the European blackbird was a timid, forest-dwelling songbird that migrated to the Mediterranean region to survive winter. Today, blackbirds are one of the most common birds found in the cities and suburbs of Europe, although their "wild" counterparts still live within forests. Urban birds seem to have adapted to city life by staying there during winter, starting to breed about three weeks earlier, and being tame enough to nest near high levels of human disturbance.

Urban environments present new opportunities and challenges for birds, and in evolutionary terms change the selection pressures that determine which behaviour is most beneficial to the individual. Urban environments tend to be warmer than nearby natural habitats because paved surfaces and buildings retain and then radiate heat, and allow birds to survive a northern winter more easily and to begin breeding earlier in spring. Deliberate feeding of birds and discarded human food provide a regular and abundant supply of food that forest birds could only dream of. On the other hand, urban environments are by definition overrun with humans, dogs, and cats, and are impoverished in terms of native biodiversity. Light and noise pollution are ubiquitous, as is chemical air and water pollution.

Jesko Partecke, from the Max Planck Institute for Ornithology in Germany, has led a series of experiments on European blackbirds to find out what behaviours seen in urban-dwelling birds are heritable. His experiments involve collecting nestlings from the nests of blackbirds in Munich versus a nearby forest, hand-rearing the nestlings under identical conditions, and subsequently allowing them to breed in aviaries. Partecke found that hand-reared males originally born to urban parents had a weaker migration activity in their first year and came into breeding condition two to three weeks earlier than birds hatched in a forest, indicating some genetic basis to these traits.

Interest in bird migration is at an all-time high, but not just because we have new gadgets and laboratory tests to reveal the amazing athletic feats of individuals or what birds are up to when they are *not* breeding. Many birds, including songbirds, shorebirds, ducks, and hawks, leave their breeding areas to survive winter. How can these birds adjust their breeding time to rapid climate change when, just a few weeks before egg-laying is to begin, they are still thousands of kilometres away in the tropics?

Epilogue
Can Birds Change with the Times?

On a recent hike in the Adirondack Mountains of New York, I soaked in the nature around me, enjoying my solitude. Hemlock and red pine trees towered overhead, the sunny spots by the trail were home to elderberry bushes loaded with bright red fruits, and a soft breeze rustled the leaves on the beech trees. My brother had kindly offered to take my kids to the beach with his little one, and for the first time in over a decade I was hiking up Ampersand Mountain without having to carry bodies, hand out water bottles and snacks, and nurse skinned knees. Despite the cool, damp air that lingered under the forest canopy, I worked up a hard sweat before I reached the summit. I had left early enough in the morning to beat the other hikers, so had the view to myself.

The panoramic view was stunning, and in one direction there was solid forest as far as I could see, a rare sight even for me. I had left my binoculars behind for practical reasons, but could hear that I shared the summit with a winter wren, a yellow-rumped warbler, a blue-headed vireo, a hermit thrush, and a junco. A group of six chimney swifts streaked back and forth twittering in a tight formation. Then a white-throated sparrow let loose its clear, ringing song, *Oh Sweet Canada, Canada, Canada,* which most Americans hear as *Old Sam Peabody, Peabody, Peabody.*

My moment of wilderness vanished when I heard a voice 20 metres away, and looked over to see a woman chatting loudly on her cell phone. Frowning and muttering to myself, I noticed the graffiti on the boulders bearing long-forgotten and uninspiring messages, such as "Dave '75." Airplane contrails crisscrossed the sky behind the swifts, and I heard the loud rumble of a truck on the highway far below my perch.

Humans have an overwhelming presence on this planet and consequently a far-reaching impact on the species around them. Our sheer numbers (over six billion and counting), the enormous scale of our footprint on the landscape, and the modernization of our agriculture, industry, and medicine, means that we are now a key player in almost every ecosystem in the world. Whether through direct habitat destruction or indirect pollution and climate change, we have dramatically altered habitats as diverse as coral reefs, tropical rainforests, and Arctic ice caps. The Acadian flycatcher arrives at its winter territory in Costa Rica to find the tropical forest replaced with a soybean field and the fairy-wren faces a long drought in southeastern Australia.

It is sobering that even birds that live in some of the most remote places on earth experience the pressures of humans. A twenty-five-year-old albatross flying for thousands of kilometres over open ocean may never return to its breeding colony if it is snagged on the hooks of a longline fishing boat. Penguins nesting on Possession Island in the south Indian Ocean cannot find enough food to stay alive because ocean waters are too warm. The ivory gull is an Arctic specialist, the avian equivalent of a polar bear, who spends its life feeding at the edges of icy seas in the far north. This species has declined by 80 percent in past decades despite its rare encounters with people because temperatures in the Arctic are rising three to five times faster than in other parts of the world.

Quetzals, which are famous for their long emerald tail plumes and live at elevations of 2,000 metres within the Poás Volcano National Park in central Costa Rica, are exposed to high levels of the pesticide endosulfan, used widely in pineapple, banana, and coffee plantations, because it drifts into the cloud forest via prevailing trade winds.

Though we typically view this damage through the lens of species extinction and biodiversity loss, this outcome is the result of the ability, or inability, of species to adapt rapidly enough to their changing world. Highly sensitive species, with specialized lifestyles and few options, may simply go extinct. Those with wider environmental tolerance may occupy human-altered landscapes but have to adapt behaviourally or genetically to new selection pressures and challenges in order to persist.

The most dramatic evolutionary redesign triggered by humans, one that has caused untold misery and economic cost, results from the rapid but unwelcome evolution of genetic resistance to our biotechnology. In evolutionary terms, pesticides and antibiotics create extremely strong selection pressure that favours the minority of individuals in pest species that happen to carry resistant genes. In 1939, Paul Müller made a discovery that changed the world, by showing that DDT killed insects; house-flies had already evolved resistance to the pesticide by the time he received his Nobel Prize in 1948. The worldwide effort to eradicate malaria by the 1960s was stopped in its tracks because mosquitoes also evolved resistance to DDT. In the 1940s most bacterial infections were treatable with penicillin, but today the vast majority of infections in hospitals are caused by serious penicillin-resistant bacteria, like *Staphylococcus,* that are also resistant to new, stronger drugs. Currently, the United States alone uses over 300 million kilograms of pesticide a year, and 25 to 50 percent of antibiotic production is used for livestock feed.

Humans have changed the evolutionary playing field, creating new selection pressures and triggering short-term evolution that, for most species, could easily be overlooked. Salmon hatcheries favour dwarf males who return from sea earlier than usual because this increases survival without putting males at a disadvantage when spawning with females. Songbirds in cities have evolved sedentary populations that breed earlier and have low stress levels to cope with the urban environment. African rainforest birds that live in cacao and coffee plantations because the rainforests were destroyed have evolved longer wings (for longer flights in open habitat), duller colouration (to better hide from predators), and shorter songs (to be heard in the windier, hotter environment). Backyard bird feeders in Britain have selected for the once-rare migration routes of European blackcaps because these seemingly off-course individuals now enjoy high winter survival and migrate earlier to the breeding grounds, getting a head start on their competitors.

These examples are evidence that birds do evolve and have

some resiliency to adapt to the modern world. If I put on my "behavioural ecologist" hat, then human-induced evolution becomes an unplanned but very interesting experiment that teaches us about how bird behaviour is shaped over time. Birds have evolved sophisticated behaviour to judge mate quality, compete with rivals using colour and song signals, and juggle the demands of family life with their own future survival. We can measure in real-time how such traits have shifted evolutionarily in recent decades—but can these subtle changes in behaviour prepare a species for surviving the next century?

Species can change rapidly if the bird in question has already evolved behavioural flexibility to cope with a wide range of natural environmental conditions. Urban areas are one of the fastest growing habitats on earth and select for individuals that can coexist with the rising tide of humanity. A comparison of urban and rural pairs of closely related bird species showed that those that have become common in urban environments have a pre-existing flexibility to thrive in a wide range of environments, predisposing them to be able to live in the concrete jungle. This flexibility includes a behavioural tendency to explore and use novel habitats, food types, or nest sites, and a physiological flexibility to reduce stress levels in urban areas.

The female robin that crouches on her bulky nest in the rhododendron beside our farmhouse has to put up with noisy kids racing down the driveway many times a day, slamming car doors, and feral cats on their nightly prowls. The male robins who bring music to the pre-dawn darkness of our Toronto suburban neighbourhood sing from rooftops, not trees. American robins have always been generalists and nest in a wide range of habitats in nature; they evolved a built-in flexibility that today allows them to breed in backyards.

Other thrushes do not have this flexibility and require a relatively untouched forest habitat. The wood thrush is roughly
the shape and size of a robin, just a little smaller, but is well
camouflaged in the forest with its rusty brown back and brown-
spotted breast. Although the male's song is conspicuous and
bold, these birds are shy and difficult to see. My students and I
have radio-tracked wood thrush, and even when we know what
tree or shrub one is hiding in we often cannot see it. The construction of houses beside forests, or the use of all-terrain vehi-
cles within a forest, can lead to wood thrush abandoning the
area altogether even if no trees are actually cut down.

The extinction of the passenger pigeon, a forest bird, seems
like a most unlikely event because less than two hundred years
ago their population numbered in the billions. These pigeons
formed enormous breeding colonies that stretched for dozens of kilometres, with ten or more nests per tree. The colony
moved from one year to the next, flying thousands of kilometres
searching for new forested areas where the beeches and oaks
had produced a bumper crop of seeds. Because they were not
tied to any one spot, pigeons always could find enough food to
fuel their amazing numbers. The eastern deciduous forests of
North America were heavily logged during the 1800s, but many
small forest patches persisted through the worst of times. Why is
it that forest thrushes, like the wood thrush, survived this habitat
destruction while the highly mobile passenger pigeons did not?

The simple, but incorrect, answer is that the pigeons were
hunted to extinction. The plump birds were easy pickings for
a hunter, who could shoot hundreds in a day. The birds were
packed in barrels and shipped by the tens of thousands to lucra-
tive markets along the east coast. Even the hunters could barely
make a dent in a breeding colony, though, and killed at most

5 percent of the entire flock in a given year. Instead it was the super-colonial behaviour of pigeons that doomed them once the forests were cleared. Nesting in enormous colonies was a behaviour as inflexible as their dependence on the seeds of forest trees. Small forest patches did not allow for large colonies, and, unfortunately, the birds refused to breed in small numbers. Deforestation happened so quickly that it was impossible for the pigeons to adapt their behaviour to the fragmented forests that had become the normal environment in eastern North America.

Flexibility can be a double-edged sword in today's world if it allows a species to live in disturbed or novel habitat, only to suffer as a result. Hooded warblers are gap specialists who, in a mature forest, set up territories in tree-fall gaps where the sunlight reaches the forest floor and spawns a thick undergrowth. These warblers readily move into recently logged forests, as long as half the trees are left standing, because this is like a giant forest gap. In one sense, it is good news that this forest bird is not so picky.

On the other hand, the magnet of the super-sized gaps leads to heavy nest failure because of a new enemy that also likes forest edges and gaps. The brown-headed cowbird is, or was, a bird of the wide, open prairies, famous because it makes a living laying its eggs in the nests of other birds, leaving the unwitting host parents to raise the egg. To make room for its egg, the cowbird steals and often eats one of the host's eggs. When the eastern forests of North America were logged, cowbirds moved east and found dozens of new species of hosts at their service. These eastern sparrows, warblers, vireos, and thrushes had rarely, if ever, encountered cowbirds in their evolutionary history, so had no tricks for evading cowbirds or even recognizing that they'd been duped.

My student Margaret Eng monitored the nesting success of hooded warblers in logged fragments in southern Ontario, and then radio-tracked the fledglings to see how many young survived after they left the nest. Cowbirds were abundant in these forest fragments, especially after logging, and almost half the warbler nests contained cowbird eggs. Most of the forest fragments that Margaret studied were eagerly occupied by hooded warblers, but nesting success was so low that their fatal attraction to partially logged areas was actually driving the population numbers down, not up. Biologists refer to these areas as "ecological traps" because the habitat looks good to the bird, but the population would be better off if individuals could somehow circumvent their instinct.

One of the biggest environmental problems facing birds, and people, is climate change, which threatens to unravel ecosystems and undermine the fundamental natural resources that human societies depend on (for example, water, food, air). Birds are good models for studying the effects of climate change because they are easily observed, their nests can be monitored closely, and they are highly sensitive to food supply. Since parental care is so demanding, most birds have evolved behavioural mechanisms to time their breeding for when food is most abundant. Small changes in temperature have a significant ripple effect at the bottom of the food chain, which alters the quantity and timing of food and potentially has a big impact on a bird's reproduction and survival.

A population of great tits has been studied in Oxford, England, since 1961 producing detailed records of when individuals started laying eggs, nesting success, and the timing and abundance of the main food, caterpillars. Great tits time their breed-

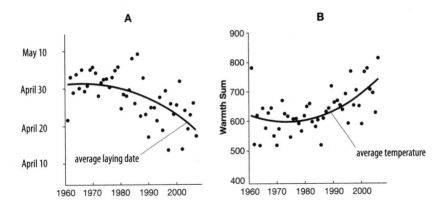

Figure 11.1. *The timing of egg-laying of great tits in Oxford, England, (A) has advanced by almost two weeks since 1960 in response to warmer spring temperatures (B), measured as the sum of maximum temperatures each day during spring. (After Charmantier et al., 2008).*

ing so their young hatch about the same time as the brief but large pulse in caterpillars, guaranteeing plenty of food for large broods of nestlings. Individuals that miss the peak in caterpillars because they laid too early or too late produce fewer offspring.

Over the past forty-seven years, the mean egg-laying date of females has advanced by about two weeks in response to higher average spring temperatures (Figure 11.1). Warmer springs mean the caterpillars appear two weeks earlier in the season, and over the decades the birds have adjusted their egg-laying to track the changing timing of the food supply. Individual females lay earlier in warm years and later in cold years. This remarkable shift was not a result of rapid evolution of genes controlling timing of breeding, but rather, the built-in plasticity in a species that is used to fine-tuning breeding to match the caterpillar supply.

Great tits in the Netherlands are not so fortunate. Females have advanced their laying dates but not fast enough to match the ever-earlier hatch dates of caterpillars. Though some females have managed the full shift, many are less flexible and lay eggs too late, only to discover they have missed the surge in caterpillar abundance that fuels nestling growth. Females that are able to lay eggs very early have a higher lifetime reproductive success than those that are behaviourally or genetically locked into normal laying times. Despite strong selection for early breeding, there is an ever-increasing mismatch between the bird's nesting period and the food supply.

The mismatch between egg-laying and insect emergence is a particularly serious problem for migratory songbirds that are still on their wintering grounds a month before nesting. By the time they arrive in spring, it may be too late to adjust laying dates. A study of pied flycatchers in the Netherlands found that this forest bird has also become out of sync with its food supply. Pied flycatcher populations have declined by 90 percent in recent decades in areas where caterpillars hatch very early, but flycatchers in forests with little shift in caterpillars are holding their own. A recent survey of one hundred migratory bird species in Europe found that species in the steepest decline are those that have not successfully shifted their egg-laying dates in response to climate change.

Examples such as these raise a warning that although birds are laying eggs earlier than ever, there is a limit to how well they can readjust their timing of breeding to match the fast pace of climate change. The changes we have witnessed so far are likely the result of built-in behavioural flexibility. Each female in a population has a genetically based tendency to breed relatively early versus late, and around this fixed optimum she can adjust her laying date a bit according to the environmental conditions she encounters each spring. The two-week shift in egg-laying that we have seen up until now could have been accomplished without genetic change because it falls within the range of possible laying dates already present in the population.

As climate change advances spring even further, it will exhaust personal flexibility and female birds will not be able to breed any earlier. This appears to have happened already for great tits and pied flycatchers in the Netherlands. Any further adaptation to climate change by the population as a whole would require genetic evolution to allow a female more flexibility or to shift a female's baseline laying time a week or two earlier. There is evidence that both plasticity and timing of breeding have a genetic basis, but evolution may not occur fast enough to prevent severe population declines.

Evolutionary shifts in egg-laying time may first require a suite of evolutionary changes in other behaviours. For long-distance migrants, climate change advances the timing of spring in their breeding areas, but the timing of spring migration itself is influenced by changing day length, built-in annual clocks, and conditions in the tropics where temperature shifts are not so pronounced. Spring migration may not advance even though migrants need to arrive earlier on their breeding grounds. In pied flycatchers, for instance, any further advancement in egg-laying date would require birds to arrive earlier in spring, but arrival dates have not advanced in recent decades.

Timing of breeding in birds is controlled at a coarse level by photoperiod, and then is fine-tuned by temperature, food supply, and social interactions with mates. Climate change affects temperature and food supply, but does not affect photoperiod. The sun rises and sets at the same time no matter how much carbon dioxide we pump into the atmosphere. The inner workings of a bird's body are locked into using photoperiod to control hormonal changes, which in turn program the growth of testes and ovaries and hence nesting and singing behaviour. This sets boundaries on what natural selection can accomplish because the intricate physiological processes that trigger breeding cannot easily be reprogrammed.

Climate change is projected to cause widespread extinction of birds and other wildlife. A model using a realistic estimate of surface warming of 2.8 degrees Celsius projects a best guess of about five hundred land bird extinctions by 2100, with another two thousand species at risk of extinction. Seabirds face even deeper problems because climate change causes a drop in food supply that persists for the whole breeding season. Shifting egg-laying dates would be of little use. For long-lived seabirds, skipping a year of breeding is an option, but climate change is expected to produce a long run of bad years, meaning a whole generation is facing failure.

～⌒

It is hard to overestimate the scope and severity of human impact on the natural world. Yet human attitudes and knowledge about the environment and our dependence on functioning ecosystems have evolved rapidly in the past century. There is hope, because humans are the ultimate example of behavioural flexibility and cultural change.

My field site in Pennsylvania is only a thirty-minute drive from Titusville, where the first oil well in the world was drilled, in 1859, by Edwin Drake. Our farmhouse, like many others in the region, is heated by our very own gas well. The museum in the Oil Creek State Park features picture after picture of bleak, denuded landscapes filled with oil derricks, streams afire with burning oil, and hillsides slick in a sea of mud. Yet, only 150 years later, I can step outside and look at a valley that is rich with oak, maple, and beech trees that provide a home to scarlet tanagers, wild turkey, and bald eagles.

The Adirondack Mountains of northeastern New York are home to the 2.5-million-hectare Adirondack Park, created in 1885, almost half of which is owned and managed by the state. New York's Constitution states that these public lands must never be developed and "shall be forever kept as wild forest lands." The idea for protecting this region first came from Verplanck Colvin, a surveyor who helped to map the wilderness. Colvin argued that the extensive logging taking place would disrupt rainfall patterns and shut down a key transportation route, the Erie Canal.

When I visited the Adirondacks as a child, in the 1960s and 1970s, the haunting cry of the loon was a rarity. Today, these birds are common; loons and many other birds at the top of the food chain fell victim to DDT poisoning in the 1950s and 1960s but have recovered since the pesticide was banned. The 1970s brought a different kind of pollution—the air that drifts northeast from coal-burning plants in Ohio, Illinois, Indiana, and Pennsylvania rises over the mountains and condenses into clouds. Nitrogen oxide and sulphur dioxide mix with moisture, which pours down as acid rain on the forests and lakes. Hundreds of lakes and ponds became too acidic to support the plants and wildlife that normally lived there. But emissions of sulphur dioxide have dropped about 40 percent since the 1970 amendment of the U.S. Clean Air Act, which was strengthened further in the 1990s, leading to a slow but encouraging recovery of Adirondack lakes that will take several more decades to complete.

A green revolution is underway, sparked largely by awareness of global climate change and the urgent need for governments and consumers to live more sustainably. For example, Ontario recently passed laws to phase out incandescent light bulbs and to ban unnecessary lawn pesticides. The U.S. Food and Drug Administration recently banned the use of the dangerous pesticide carbofuran, and no longer allows it on imported foods. In June 2009, the U.S. House of Representatives passed a historic climate change bill, the American Clean Energy and Security Act, which aims to cut U.S. production of greenhouse gases by 17 percent by 2020, and 83 percent by 2050.

Nature is remarkably resilient, and has shown that species can bounce back if given enough time and if we rein in our heavy-handed use of resources and chemicals.

~⌒

I had hoped to see a new bird, the elusive Bicknell's thrush, for my "life list," all the species I have seen since I started keeping track, while up on Ampersand Mountain. This bird was first discovered in 1881, but is so similar to the grey-cheeked thrush that it was not named a distinct species until 1995. Shy and reclusive, like most thrushes, the Bicknell's takes this one step further in both its preference for remote mountaintops and its reluctance to be seen by even the most persistent birder.

By virtue of its rarity, the Bicknell's thrush is highly threatened. Populations are scattered and isolated, and are vulnerable to deforestation within its small breeding range in northeastern North America and on the few Caribbean islands where it spends the winter. It is especially sensitive to climate warming that erodes the cold, alpine forests, and pollutants such as mercury condense at high elevation and have been found in alarming concentrations in the Bicknell's thrush. I had a special interest in this bird for a fun reason: its mating system is intriguing. Males do not defend a breeding territory, unlike other thrushes, and a brood of nestlings may be fed by up to four different males, only some of whom have a genetic stake in the young.

Since I could not count on seeing my quarry, I used my ears. The song of the Bicknell's thrush is a buzzy, spiralling whistle that ends with a noticeable rising note. Though I did pick out the familiar flute-like song of a hermit thrush, and the harsh bark of a veery, no mysterious songs leapt out from the stunted spruce trees near the summit. As I picked my way down the steep trail, sometimes hanging onto the roots of trees, I heard a high-pitched and sharp call *beer*. I stopped in my tracks and held my breath, watching for the slightest movement. Five minutes later I reluctantly continued on my way.

When I got home later that day I listened to the Bicknell's thrush call notes on my iPod, but couldn't be sure that what I'd heard was a perfect match. I'll return next summer and stop at the same place, hoping the elusive bird will be back too.

ACKNOWLEDGMENTS

I didn't consider myself a "writer" until a few years ago when *Silence of the Songbirds,* in which I chronicled the alarming songbird declines in North America, was published. This may come as a surprise, because I have published almost one hundred articles in scientific journals since 1984, plus an academic book on the behaviour of tropical birds. The difference between those scholarly works and *Silence of the Songbirds,* however, lies in the audience. Here for the first time I was writing for naturalists, bird watchers, and neighbours down the street. I thank Jim Gifford at Harper Collins for inviting me to write my first trade book and for drawing me into this very rewarding adventure.

The Private Lives of Birds was written at my kitchen table, at the desk in my bedroom, and in an old farmhouse in Pennsylvania, so I must thank my husband and children yet again for their patience in putting up with my apparent daydreaming and occasional indifference to household chores. Jim Gifford, along with Jackie Johnson of Walker & Co., were excellent editors whom I enjoyed working with. My grammar reflects a long career in the sciences, and the final product was polished with careful copy editing by Allyson Latta.

Finally, I should thank the birds who put up with my interruptions and heavy demands during the course of my research. My discoveries would not have been possible without their cooperation in my mist netting, song playbacks, radio-tracking, plumage dyeing, blood sampling, geo-locating, and so on. Proceeds from *The Private Lives of Birds* will be used to fund bird conservation research.

SOURCES

INTRODUCTION

Evans, M.L., B.J.M. Stutchbury, and B.E. Woolfenden. "Off-Territory Forays and the Genetic Mating System of the Wood Thrush." *The Auk* 125 (2008): 67–75.

Klatt, P.H., B.J.M. Stutchbury, and M.L. Evans. "Incubation Feeding in the Scarlet Tanager: a Removal Experiment." *Journal of Field Ornithology* 79 (2008): 1–10.

Moore, L., B.J.M. Stutchbury, D. Burke, and K. Elliot. "Effects of Forest Management on Post-Fledging Survival of the Rose-Breasted Grosbeak." *The Auk* 134 (2010): 185–194.

Robertson, R.J., and B.J. Stutchbury. "Experimental Evidence for Sexually Selected Infanticide in Tree Swallows." *Animal Behaviour* 36 (1988): 749–753.

Robertson, R.J., H.L. Gibbs, and B.J. Stutchbury. "Spitefulness, Altruism, and the Cost of Aggression: Evidence against Superterritoriality in Tree Swallows." *The Condor* 88 (1986): 123–124.

Rush, S.A., and B.J.M. Stutchbury. "Survival of Fledgling Hooded Warblers in Large and Small Forest Fragments." *The Auk* 125 (2008): 183–191.

1: PHILANDERING FLYCATCHERS

Double, M.C., and A. Cockburn. "Pre-Dawn Infidelity: Females Control Extra-Pair Mating in Superb Fairy-Wrens." Proceedings of the Royal Society of London B267 (2000): 465–470.

Drent, P.J., K. van Oers, and A.J. van Noordwijk. "Realized Heritability of Personalities in the Great Tit (*Parus major*)." Proceedings of the Royal Society of London B270 (2003): 45–51.

Dunn, P.O., and A. Cockburn. "Extrapair Mate Choice and Honest
Signaling in Cooperatively Breeding Superb Fairy-Wrens."
Evolution 53 (1999): 938–946.

Freeman-Gallant, C.R., N.T. Wheelwright, K.E. Meiklejohn, and S.V.
Sollecito. "Genetic Similarity, Extrapair Paternity, and Offspring
Quality in Savannah Sparrows (*Passerculus sandwichensis*)."
Behavioral Ecology 17 (2006): 952–958.

Hung, S., S.A. Tarof, and B.J.M. Stutchbury. "Extra-Pair Mating
Tactics and Vocal Behaviour in Female Acadian Flycatchers,
Empidonax virescens." *The Condor* 111 (2009): 653–661.

Perlut, N.G., C.R. Freeman-Gallant, A.M. Strong, T.M. Donovan,
C.W. Kilpatrick, and N.J. Zalik. "Agricultural Management Affects
Evolutionary Processes in a Migratory Songbird." *Molecular
Ecology* 17 (2008): 1248–1255.

Pulido, F., P. Berthold, G. Mohr, and U. Querner. "Heritability of
the Timing of Autumn Migration in a Natural Bird Population."
Proceedings of the Royal Society of London B268 (2001):
953–959.

Tobias, J.A., and N. Seddon. "Female Begging in European Robins:
Do Neighbors Eavesdrop for Extrapair Copulations?" *Behavioral
Ecology* 13 (2002): 637–642.

Wagner, R.H. "Extra-Pair Copulations in a Lek: The Secondary
Mating System of Monogamous Razorbills." *Behavioral Ecology and
Sociobiology* 31 (1992): 63–71.

Woolfenden, B.E., B.J.M. Stutchbury, and E.S. Morton. "Extra-Pair
Fertilizations in the Acadian Flycatcher: Males Obtain EPFs with
Distant Females." *Animal Behaviour* 69 (2005): 921–929.

2: Monogamy in a Tropical Paradise

Evans, M.L., B.J.M. Stutchbury, and B.E. Woolfenden. "Off-Territory
Forays and the Genetic Mating System of the Wood Thrush." *The
Auk* 125 (2008): 67–75.

Fleischer, R.C., C.L. Tarr, E.S. Morton, A. Sangmeister, and K.C. Derrickson. "Mating System of the Dusky Antbird, a Tropical Passerine, as Assessed by DNA Fingerprinting." *The Condor* 99 (1997): 512–514.

Gill, S.A., M.J. Vonhof, B.J.M. Stutchbury, E.S. Morton, and J.S. Quinn. "Does Duetting during the Dawn Chorus Announce Female Fertility?" *Behavioral Ecology and Sociobiology* 57 (2005): 557–565.

Moore, O., B.J.M. Stutchbury, and J.S. Quinn. "Extra-Pair Mating System of an Asynchronously Breeding Tropical Songbird, the Mangrove Swallow." *The Auk* 116 (1999): 1039–1046.

Stutchbury, B.J.M., and E.S. Morton. *Behavioral Ecology of Tropical Songbirds*. London: Academic Press, 2001.

———. "Recent Advances in the Behavioral Ecology of Tropical Birds." *Wilson Journal of Ornithology* 120 (2008): 26–37.

Stutchbury, B.J.M., E.S. Morton, and W.H. Piper. "Extra-Pair Mating System of a Synchronously Breeding Tropical Songbird." *Journal of Avian Biology* 29 (1998): 72–78.

3: Finicky Females

Aguilar, T.M., R. Maia, E.S.A. Santos, and R.H. Macedo. "Parasite Levels in Blue-Black Grassquits Correlate with Male Displays But Not Female Mate Preference." *Behavioral Ecology* 19 (2008): 292–301.

Gwinner, H., and H. Schwabl. "Evidence for Sexy Sons in European Starlings (*Sturnus vulgaris*)." *Behavioral Ecology and Sociobiology* 58 (2005): 375–382.

Hill, G.E. *A Red Bird in a Brown Bag: The Function and Evolution of Ornamental Plumage Coloration in the House Finch*. New York: Oxford University Press, 2002.

Hill, G.E., and K.L. Farmer. "Carotenoid-Based Plumage Coloration Predicts Resistance to a Novel Parasite in the House Finch." *Naturwissenschaften* 92 (2005): 30–34.

Préault, M., O. Chastel, F. Cézilly, and B. Faivre. "Male Bill Colour and Age Are Associated with Parental Abilities and Breeding Performance in Blackbirds." *Behavioral Ecology and Sociobiology* 58 (2005): 497–505.

Siefferman, L., and G.E. Hill. "Structural and Melanin Plumage Coloration Indicate Parental Effort and Reproductive Success in Male Eastern Bluebirds." *Behavioral Ecology* 14 (2003): 855–861.

Smith, T.B., B. Milá, G.F. Grether, H. Slabbekoorn, I. Sepil, W. Buermann, S. Saatchi, and J.P. Pollinger. "Evolutionary Consequences of Human Disturbance in a Rainforest Bird Species from Central Africa." *Molecular Ecology* 17 (2008): 58–71.

Torres, R., and A. Velando. "A Dynamic Trait Affects Continuous Pair Assessment in the Blue-Footed Booby, *Sula nebouxii*." *Behavioral Ecology and Sociobiology* 55 (2003): 65–72.

Velando, A., R. Torres, and I. Espinosa. "Male Coloration and Chick Condition in Blue-Footed Booby: A Cross-Fostering Experiment." *Behavioral Ecology and Sociobiology* 58 (2005): 175–180.

4: Avian Operas

Barnett, C.A., and J.V. Briskie. "Energetic State and the Performance of Dawn Chorus in Silvereyes (*Zosterops lateralis*)." *Behavioral Ecology and Sociobiology* 61 (2007): 579–587.

Chiver, I., B.J.M. Stutchbury, and E.S. Morton. "Female Foray Behaviour Correlates with Male Song and Paternity in a Socially Monogamous Bird." *Behavioral Ecology and Sociobiology* 62 (2008): 1981–1990.

Duffy, D.L., and G.F. Ball. "Song Predicts Immunocompetence in Male European Starlings *Sturnus vulgaris*." Proceedings of the Royal Society of London B269 (2002): 847–852.

Gorissen, L., T. Snoeijs, E. Van Duyse, and M. Eens. "Heavy Metal Pollution Affects Dawn Singing Behaviour in a Small Passerine Bird." *Oecologia* 145 (2005): 504–509.

Iwaniuk, A.N., D.T. Koperski, K.M. Cheng, J.E. Elliott, L.K. Smith, L.K. Wilson, and D.R.W. Wylie. "The Effects of Environmental Exposure to DDT on the Brain of a Songbird: Changes in Structures Associated with Mating and Song." *Behavioural Brain Research* 173 (2006): 1–10.

Markman, S., S. Leitner, C. Catchpole, S. Barnsley, C.T. Müller, D. Pascoe, and K.L. Buchanan. "Pollutants Increase Song Complexity and the Volume of the Brain Area HVC in a Songbird." 2008. PLoS ONE 3(2): e1674. doi:10.1371/journal.pone.0001674.

Morton, E.S. "Ecological Sources of Selection on Avian Sounds." *The American Naturalist* 109 (1975): 17–34.

Morton, E.S., J. Howlett, N.C. Kopysh, and I. Chiver. "Song Ranging by Incubating Male Blue-Headed Vireos: The Importance of Song Representation in Repertoires and Implications for Song Delivery Patterns and Local/Foreign Dialect Discrimination." *Journal of Field Ornithology* 77 (2006): 291–301.

Nowicki, S., W. Searcy, and S. Peters. "Brain Development, Song Learning and Mate Choice in Birds: A Review and Experimental Test of the Nutritional Stress Hypothesis." *Journal of Comparative Physiology* 188 (2004): 1003–1014.

Slabbekoorn, H., and M. Peet. "Birds Sing at a Higher Pitch in Urban Noise." *Nature* 424 (2003): 267.

Slabbekoorn, H., and E.A.P. Ripmeester. "Birdsong and Anthropogenic Noise: Implications and Applications for Conservation." *Molecular Ecology* 17 (2008): 72–83.

Soma, M., M. Hiraiwa-Hasegawa, and K. Okanoya. "Early Ontogenetic Effects on Song Quality in the Bengalese Finch (*Lonchura striata* var. *domestica*): Laying Order, Sibling Competition, and Song Syntax." *Behavioral Ecology and Sociobiology* 63 (2009): 363–370.

Spencer, K.A., K.L. Buchanan, S. Leitner, A.R. Goldsmith, and C.K. Catchpole. "Parasites Affect Song Complexity and Neural Development in a Songbird." Proceedings of the Royal Society of London B272 (2005): 2037–2043.

5: 'TIL DEATH DO US PART

Awkerman, J.A., K.P. Huyvaert, J. Mangel, J.A. Shigueto, and D.J. Anderson. "Incidental and Intentional Catch Threatens Galapagos Waved Albatross." *Biological Conservation* 133 (2006): 483–489.

Beissinger, S.R. "Experimental Brood Manipulations and the Monoparental Threshold in Snail Kites." *The American Naturalist* 136 (1990): 20–38.

BirdLife International's "Save the Albatross" Campaign. http://www. savethealbatross.net

Gill, S.A., and B.J.M. Stutchbury. "Long-Term Mate and Territory Fidelity in Neotropical Buff-Breasted Wrens (*Thryothorus leucotis*)." *Behavioral Ecology and Sociobiology* 61 (2006): 245–253.

Greenberg, R., and J. Gradwohl. "Territoriality, Adult Survival, and Dispersal in the Checker-Throated Antwren in Panama." *Journal of Avian Biology* 28 (1997): 103–110.

Griggio, M., G. Matessi, and A. Pilastro. "Should I Stay or Should I Go? Female Brood Desertion and Male Counterstrategy in Rock Sparrows." *Behavioral Ecology* 16 (2005): 435–441.

Griggio, M., and A. Pilastro. "Sexual Conflict over Parental Care in a Species with Female and Male Brood Desertion." *Animal Behaviour* 74 (2007): 779–785.

Heg, D., L.W. Bruinzeel, and B.J. Ens. "Fitness Consequences of Divorce in the Oystercatcher, *Haematopus ostralegus*." *Animal Behaviour* 66 (2003): 175–184.

Jeschke, J.M., S. Wanless, M.P. Harris, and H. Kokko. "How Partnerships End in Guillemots *Uria aalge*: Chance Events, Adaptive Change, or Forced Divorce?" *Behavioral Ecology* 18 (2007): 460–466.

Morton, E.S., K.C. Derrickson, and B.J.M. Stutchbury. "Territory Switching Behavior in a Sedentary Tropical Passerine, the Dusky Antbird." *Behavioral Ecology* 6 (2000): 648–653.

Van de Pol, M., D. Heg, L.W. Bruinzeel, B. Kuijper, and S. Verhulst. "Experimental Evidence for a Causal Effect of Pair-Bond Duration on Reproductive Performance in Oystercatchers (*Haematopus ostralegus*)." *Behavioral Ecology* 17 (2006): 982–991.

Van Dijk, R.E., I. Szentirmai, J. Komdeur, and T. Szekely. "Sexual Conflict over Parental Care in Penduline Tits (*Remiz pendulinus*): The Process of Clutch Desertion." *Ibis* 149 (2007): 530–534.

Xavier, J.C., P.N. Trathan, J.P. Croxall, A.G. Wood, G. Podesta, and P.G. Rodhouse. "Foraging Ecology and Interactions with Fisheries of Wandering Albatrosses (*Diomedea exulans*) Breeding at South Georgia." *Fisheries Oceanography* 13 (2004): 324–344.

6: Your Turn or Mine?

Addison, B., A.S. Kitaysky, and J.M. Hipfner. "Sex Allocation in a Monomorphic Seabird with a Single-Egg Clutch: Test of the Environment, Mate Quality, and Female Condition Hypotheses." *Behavioral Ecology and Sociobiology* 63 (2008): 135–141.

Bonier, F., P.R. Martin, and J.C. Wingfield. Maternal Corticosteroids Influence Primary Offspring Sex Ratio in a Free-Ranging Passerine Bird. *Behavioral Ecology* 18 (2007): 1045–1050.

Cameron-MacMillan, M.L., C.J. Walsh, S.I. Wilhelm, and A.E. Storey. "Male Chicks Are More Costly to Rear Than Females in a Monogamous Seabird, the Common Murre." *Behavioral Ecology* 18 (2007): 81–85.

Chiver, I., E.S. Morton, and B.J.M. Stutchbury. "Male Blue-Headed Vireos Delay Territorial Defense While on the Nest Incubating." *Animal Behaviour* 73 (2007): 143–148.

Clouta, M.N., G.P. Elliott, and B.C. Robertson. "Effects of Supplementary Feeding on the Offspring Sex Ratio of Kakapo: A Dilemma for the Conservation of a Polygynous Parrot." *Biological Conservation* 107 (2002): 13–18.

Covas, R., A. Dalecky, A. Caizergues, and C. Doutrelant. "Kin Associations and Direct vs Indirect Fitness Benefits in Colonial Cooperatively Breeding Sociable Weavers (*Philetairus socius*)." *Behavioral Ecology and Sociobiology* 60 (2006): 323–331.

Covas, R., C. Doutrelant, and M.A. du Plessis. "Experimental Evidence of a Link between Breeding Conditions and the Decision to Breed or to Help in a Colonial Cooperative Bird." Proceedings of the Royal Society of London B271 (2004): 827–832.

Covas, R., and M.A. du Plessis. "The Effect of Helpers on Artificially Increased Brood Size in Sociable Weavers (*Philetairus socius*)." *Behavioral Ecology and Sociobiology* 57 (2005): 631–636.

Ekstrom, J.M.M., T. Burke, L. Randrianaina, and T.R. Birkhead. "Unusual Sex Roles in a Highly Promiscuous Parrot: The Greater Vasa Parrot, *Caracopsis vasa*." *Ibis* 149 (2007): 313–320.

Komdeur, J., S. Daan, J. Tinbergen, and C. Mateman. "Extreme Adaptive Modification in Sex Ratio of the Seychelles Warbler Eggs." *Nature* 385 (1997): 522–525.

Morton, E.S., B.J.M. Stutchbury, J.S. Howlett, and W.H. Piper. "Genetic Monogamy, Breeding Synchrony and Male Parental Care in the Blue-Headed Vireo." *Behavioral Ecology* 9 (1998): 515–524.

Pitcher, T.E., and B.J.M. Stutchbury. "Extraterritorial Forays and Male Parental Care in Hooded Warblers." *Animal Behaviour* 59 (2000): 1261–1269.

Russell, A.F., N.E. Langmore, A. Cockburn, L.B. Astheimer, and R.M. Kilner. "Reduced Egg Investment Can Conceal Helper Effects in Cooperatively Breeding Birds." *Science* 317 (2007): 941–944.

Sharp, S.P, A. McGowan, M.J. Wood, and B.J. Hatchwell. "Learned Kin Recognition Cues in a Social Bird." *Nature* 434 (2005): 1127–1130.

Stutchbury, B.J., J.M. Rhymer, and E.S. Morton. "Extra-Pair Paternity in the Hooded Warbler." *Behavioral Ecology* 5 (1994): 384–392.

7: Empty Nest

Eikenaar, C., D.S. Richardson, L. Brouwer, and J. Komdeur. "Parent
Presence, Delayed Dispersal, and Territory Acquisition in the
Seychelles Warbler." *Behavioral Ecology* 18 (2007): 874–879.

Griesser, M., M. Nystrand, S. Eggers, and J. Ekman. "Social
Constraints Limit Dispersal and Settlement Decisions in a Group-
Living Bird Species." *Behavioral Ecology* 19 (2008): 317–324.

Robertson, R.J., and B.J. Stutchbury. "Experimental Evidence for
Sexually Selected Infanticide in Tree Swallows." *Animal Behaviour*
36 (1988): 749–753.

Smith, S.M. "Flock Switching in Chickadees: Why Be a Winter
Floater?" *The American Naturalist* 123 (1984): 81–98.

Stutchbury, B.J. "Floater Behavior and Territory Acquisition in Male
Purple Martins." *Animal Behaviour* 42 (1991): 435–443.

———. "The Adaptive Significance of Male Subadult Plumage in
Purple Martins: Plumage Dyeing Experiments." *Behavioral Ecology
and Sociobiology* 29 (1991): 297–306.

Stutchbury, B.J., and R.J. Robertson. "Signalling Subordinate and
Female Status: Two Hypotheses for the Adaptive Significance
of Subadult Plumage in Female Tree Swallows." *The Auk* 104
(1987): 717–723.

8: Fight or Flight

Amrhein, V., and N. Erne. "Dawn Singing Reflects Past Territorial
Challenges in the Winter Wren." *Animal Behaviour* 71 (2006):
1075–1080.

Chaine, A.S., and B.E. Lyon. "Intrasexual Selection on Multiple
Plumage Ornaments in the Lark Bunting." *Animal Behaviour* 76
(2008): 657–667.

Fedy, B.C., and B.J.M. Stutchbury. "Territory Defence in Tropical Birds: Are Females as Aggressive as Males?" *Behavioral Ecology and Sociobiology* 58 (2005): 414–422.

———. "Testosterone does not Increase in Response to Conspecific Challenges in the White-bellied Antbird (*Myrmeciza longipes*), a Resident Tropical Passerine." *The Auk* 123 (2006): 61–66.

Heinsohn, R. "The Ecological Basis of Unusual Sex Roles in Reverse-Dichromatic Eclectus Parrots." *Animal Behaviour* 76 (2008): 97–103.

Heinsohn, R., S. Legge, and J.A. Endler. "Extreme Reversed Sexual Dichromatism in a Bird without Sex Role Reversal." *Science* 309 (2005): 617–619.

Hung, S., S.A. Tarof, and B.J.M. Stutchbury. "Extra-Pair Mating Tactics and Vocal Behaviour in Female Acadian Flycatchers, *Empidonax virescens*." *The Condor* 111 (2009): 653–661

McGlothlin, J.W., J.M. Jawor, T.J. Greives, J.M. Castro, J.L. Phillips, and E.D. Ketterson. "Hormones and Honest Signals: Males with Larger Ornaments Elevate Testosterone More When Challenged." *Journal of Evolutionary Biology* 21 (2007): 39–48.

Pärn, H., K.M. Lindström, M. Sandell, and T. Amundsen. "Female Aggressive Response and Hormonal Correlates—An Intrusion Experiment in a Free-Living Passerine." *Behavioral Ecology and Sociobiology* 62 (2008): 1665–1677.

Schmidt, R., V. Amrhein, H.P. Kunc, and M. Naguib. "The Day After: Effects of Vocal Interactions on Territory Defence in Nightingales." *Journal of Animal Ecology* 76 (2007): 168–173.

Schmidt, R., H.P. Kunc, V. Amrhein, and M. Naguib. "Aggressive Responses to Broadband Trills Are Related to Subsequent Pairing Success in Nightingales." *Behavioral Ecology* 19 (2008): 635–641.

Stutchbury, B.J. "Competition for Winter Territories in a Neotropical Migrant Songbird: The Role of Age, Sex, and Color." *The Auk* 111 (1994): 63–69.

van Dongen, W.F.D., and R.A. Mulder. "Relative Importance of Multiple Plumage Ornaments as Status Signals in Golden Whistlers (*Pachycephala pectoralis*)." *Behavioral Ecology and Sociobiology* 62 (2007): 77–86.

9: BIRD CITIES

Augustin, J., D. Blomquist, T. Szép, Z.D. Szabó, and R.H. Wagner. "No Evidence of Genetic Benefits from Extra-Pair Fertilizations in Female Sand Martins (*Riparia riparia*)." *Journal of Ornithology* 148 (2007): 189–198.

Brown, C.R., and M.B. Brown. *Coloniality in the Cliff Swallow: The Effect of Group Size on Social Behavior*. Chicago: University of Chicago Press, 1996.

———. "Testis Size Increases with Colony Size in Cliff Swallows." *Behavioral Ecology* 14 (2003): 569–575.

Brown, C.R., N. Komar, S.B. Quick, R.A. Sethi, N.A. Panella, M.B. Brown, and M. Pfeffer. "Arbovirus Infection Increases with Group Size." Proceedings of the Royal Society of London, B268 (2001): 1833–1840.

Cubie, D. "Seabird Signals." *National Wildlife Magazine* (Aug./Sept. 2008): 24–32.

Danchin, E., T. Boulinier, and M. Massot. "Conspecific Reproductive Success and Breeding Habitat Selection: Implications for the Study of Coloniality." *Ecology* 79 (1998): 2415–2428.

Le Bohec, C.J.M. Durant, M. Gauthier-Clerc, N.C. Stenseth, Y. Park, R. Pradel, D. Grémillet, J. Gendner, and Y. Le Maho. "King Penguin Population Threatened by Southern Ocean Warming." Proceedings of the National Academy of Sciences 105 (2008): 2493–2497.

Lee, D.L., N. Nur, and W.J. Sydeman. "Climate and Demography of the Planktivorous Cassin's Auklet, *Ptychoramphus aleuticus,* off Northern California: Implications for Population Change." *Journal of Animal Ecology* 76 (2007): 337–347.

Parker, M.W., S.W. Kress, R.T. Golightly, H.R. Carter, E.B. Parsons, S.E. Schubel, J.A. Boyce, G.J. McChesney, and S.M. Wisely. "Assessment of Social Attraction Techniques Used to Restore a Common Murre Colony in Central California." *Waterbirds* 30 (2007): 17–28.

Roby, D.D., K. Collis, D.E. Lyons, D.P. Craig, J.Y. Adkins, A.M. Myers, and R.M. Suryan. "Effects of Colony Relocation on Diet and Productivity of Caspian Terns." *Journal of Wildlife Management* 66 (2002): 662–673.

Sergio, F., and V. Penteriani. "Public Information and Territory Establishment in a Loosely Colonial Raptor." *Ecology* 86 (2005): 340–346.

Serrano, S., D. Oro, E. Ursú, and J.L. Tella. "Colony Size Selection Determines Adult Survival and Dispersal Preferences: Allee Effects in a Colonial Bird." *The American Naturalist* 166 (2005): E22–E31.

Smith, L.C., S.A. Raouf, M.B. Brown, J.C. Wingfield, and C.R. Brown. "Testosterone and Group Size in Cliff Swallows: Testing the 'Challenge Hypothesis' in a Colonial Bird." *Hormones and Behavior* 47 (2005): 76–82.

Sovada, M.A., P.J. Pietz, K.A. Converse, D.T. King, E.K. Hofmeister, and P. Scherr. "Impact of West Nile Virus and Other Mortality Factors on American White Pelicans at Breeding Colonies in the Northern Plains of North America." *Biological Conservation* 141 (2008): 1021–1031.

Stutchbury, B.J. "Evidence that Bank Swallow Colonies Do Not Function as Information Centers." *The Condor* 90 (1988): 953–955.

Wagner, R.H., E. Danchin, T. Boulinier, and F. Helfenstein. "Colonies as Byproducts of Commodity Selection." *Behavioral Ecology* 11 (2000): 572–573.

10: FREQUENT FLIERS

Bearhop, S., W. Fiedler, R.W. Furness, S.C. Votier, S. Waldron, J. Newton, G.J., Bowen, P. Berthold, and K. Farnsworth. "Assortative Mating as a Mechanism for Rapid Evolution of a Migratory Divide." *Science* 310 (2005): 502–504.

Berthold, P., A.J. Helbig, G. Mohr, and U. Querner. "Rapid Microevolution of Migratory Behavior in a Wild Bird Species." *Nature* 360 (1992): 668–670.

Brown, C.R. "Purple Martin *(Progne subis).*" In *The Birds of North America.* No. 287, edited by A. Poole and F. Gill. Philadelphia: The Academy of Natural Sciences, and Washington: The American Ornithologists' Union, 1997.

Evans Ogden, L.G. and B.J.M. Stutchbury. "Constraints on Double Brooding in a Neotropical Migrant, the Hooded Warbler." *The Condor* 98 (1996): 736–744.

Gustaffson, L., and T. Pärt. "Acceleration of Senescence in the Collared Flycatcher *Ficedula albicollis* by Reproductive Costs." *Nature* 347 (1990): 279–281.

Partecke, J., and E. Gwinner. "Increased Sedentariness in European Blackbirds Following Urbanization: A Consequence of Local Adaptation?" *Ecology* 88 (2007): 882–890.

Partecke, J., I. Schwabl, and E. Gwinner. "Stress and the City: Urbanization and Its Effects on the Stress Physiology in European Blackbirds." *Ecology* 87 (2006): 1945–1952.

Pulido, F., P. Berthold, G. Mohr, and U. Querner. "Heritability of the Timing of Autumn Migration in a Natural Bird Population." Proceedings of the Royal Society of London B268 (2001): 953–959.

Stutchbury, B.J.M., J.R. Hill III, P.M. Kramer, S.A. Rush, and S.A. Tarof. "Sex and Age-Specific Annual Survival in a Neotropical Migratory Songbird, the Purple Martin *(Progne subis).*" *Auk* 126 (2009): 278–287.

Stutchbury, B.J.M., S.A. Tarof, T. Done, E. Gow, P.M. Kramer, J. Tautin, J.W. Fox, and V. Afanasyev. "Tracking Long-Distance Songbird Migration Using Geolocators." *Science* 323 (2009): 896.

Wikelski, M., R.W. Kays, N.J. Kasdin, K. Thorup, J.A. Smith, and G.W. Swenson, Jr. "Going Wild: What a Global Small-Animal Tracking System Could Do for Experimental Biologists." *Journal of Experimental Biology* 210 (2007): 181–186.

EPILOGUE

Bonier, F., P.R. Martin, K.S. Sheldon, J.P. Jensen, S.L. Foltz, and J.C. Wingfield. "Sex-Specific Consequences of Life in the City." *Behavioral Ecology* 18 (2007): 121–129.

Bonier, F., P.R. Martin, and J.C. Wingfield. "Urban Birds Have Broader Environmental Tolerance." *Biology Letters* 3 (2007): 670–673.

Both, C., S. Bouwhuis, C.M. Lessells, and M.E. Visser. "Climate Change and Population Declines in a Long-Distance Migratory Bird." *Nature* 441 (2006): 81–83.

Both, C., and M.E. Visser. "Adjustment to Climate Change Is Constrained by Arrival Date in a Long-Distance Migrant Bird." *Nature* 411 (2001): 296–298.

Charmantier, A.R., H. McCleery, L.R. Cole, C. Perrins, L.E.B. Kruuk, and B.C. Sheldon. "Adaptive Phenotypic Plasticity in Response to Climate Change in a Wild Bird Population." *Science* 320 (2008): 800–803.

Daly, G.L., Y.D. Lei, C. Teixeira, D.C.G. Muir, L.E. Castillo, and F. Wania. "Accumulation of Current-Use Pesticides in Neotropical Montane Forests." *Environmental Science and Technology* 41 (2007): 1118–1123.

Driscoll, C.T., G.B. Lawrence, A.J. Bulger, T.J. Butler, C.S. Cronan, C. Eagar, K.F. Labert, G.E., Likens, J.L. Stoddard, and K.C. Weathers. "Acidic Deposition in the Northeastern United States: Sources

and Inputs, Ecosystem Effects, and Management Strategies."
BioScience 51 (2001): 180–198.

Goetz, J.E., K.P. McFarland, and C.C. Rimmer. "Multiple Paternity
and Multiple Male Feeders in Bicknell's Thrush (*Catharus
bicknelli*)." *The Auk* 120 (2003): 1044–1053.

Møller, A.P., D. Rubolini, and E. Lehikoinen. "Populations of
Migratory Bird Species That Did Not Show a Phenological
Response to Climate Change Are Declining." Proceedings of the
National Academy of Sciences 105 (2008): 16195–16200.

Nussey, D.H., E. Postma, P. Gienapp, M.E. Visser. "Selection on
Heritable Phenotypic Plasticity in a Wild Bird Population." *Science*
310 (2005): 304–306.

Palumbi, S.R. "Humans as the World's Greatest Evolutionary Force."
Science 293 (2001): 1786–1790.

Rimmer, C.C., K.P. McFarland, D.C. Evers, E.K. Miller, Y. Aubry, D.
Busby, and R.J. Taylor. "Mercury Concentrations in Bicknell's
Thrush and Other Insectivorous Passerines in Montane Forests of
Northeastern North America." *Ecotoxicology* 14 (2005): 223–240.

Sekercioglu, C.H., S.H. Schneider, J.P. Fay, and S.R. Loarie.
"Climate Change, Elevational Range Shifts, and Bird Extinctions."
Conservation Biology 22 (2008): 140–150.

Visser, M.E. "Keeping Up with a Warming World: Assessing the Rate
of Adaptation to Climate Change." Proceedings of the Royal
Society of London B275 (2008): 649–659.

Vyn, G. "The Allure of the Ivory Gull." *Living Bird* 28 (2009): 31–36.

INDEX

*Page numbers in **boldface** refer to material,
information and data featured in figures and illustrations.*

A Note on the Author

Bridget Stutchbury completed her PhD at Yale University, was a research associate at the Smithsonian Institution, and is now a professor of biology at York University in Toronto, where she holds a Canada Research Chair in ecology and conservation biology. Recognized as an international birding expert, she is affiliated with more than a dozen organizations seeking to preserve bird habitats, including the World Wildlife Fund. Stutchbury is the author of *Silence of the Songbirds*, and lives in Woodbridge, Ontario, and in Cambridge Springs, Pennsylvania. Visit her Web site at www.theprivatelivesofbirds. com.

Available online and at your local bookseller,
Bridget Stutchbury's eye-opening book:

Silence of the Songbirds
Foreword by John Flicker, president of the National
Audubon Society
$16.00 • Paperback
ISBN: 978-0-8027-1691-0

"Page after page here recounts surprising details taken from the lives of birds pursuing their destinies within the grand yet heartbreaking drama of 21st-century migration."—**Frank Graham Jr.,** *Audubon*

"Wonderfully informative of beautiful little things. This book is a must-read for anyone whose heart has thrilled to the song of a bird."
—**Tim Flannery, author of** *The Weather Makers*

Songbirds are a vital part of our ecosystem. Without them, our forests would face more insect infestations and our trees, flowers, and gardens would lose a crucial link in their reproductive cycle. Yet migratory songbirds, including wood thrushes, bobolinks, and Eastern kingbirds, are disappearing at a frightening rate. By some estimates, we may already have lost almost half the number of songbirds that filled the skies only forty years ago.

Following migratory birds on their six-thousand-mile journey from the tropics to North America, renowned biologist Bridget Stutchbury leads us on an ecological field trip to explore firsthand the lives of songbirds and the major threats they face, and the things each one of us can do to help save them. As *Silence of the Songbirds* shows, we ultimately protect ourselves and our children by taking steps to save songbirds.

www.yorku.ca/bstutch/research.htm